PSEUDO SCIENCE

AN AMUSING HISTORY OF CRACKPOT IDEAS AND WHY WE LOVE THEM

LYDIA KANG, MD

NATE PEDERSEN

WORKMAN PUBLISHING • NEW YORK

Copyright © 2024 by Lydia Kang and Nate Pedersen

Hachette Book Group supports the right to free expression and the value of copyright. The purpose of copyright is to encourage writers and artists to produce the creative works that enrich our culture.

The scanning, uploading, and distribution of this book without permission is a theft of the author's intellectual property. If you would like permission to use material from the book (other than for review purposes), please contact permissions@hbgusa.com. Thank you for your support of the author's rights.

Workman
Workman Publishing
Hachette Book Group, Inc.
1290 Avenue of the Americas
New York, NY 10104
workman.com

Workman is an imprint of Workman Publishing, a division of Hachette Book Group, Inc. The Workman name and logo are registered trademarks of Hachette Book Group, Inc.

Design by Galen Smith
Cover credit: (Alien) Mixzer (Generated with AI)/Freepik
Photo credit information is on page 308.

The publisher is not responsible for websites (or their content) that are not owned by the publisher.

The Hachette Speakers Bureau provides a wide range of authors for speaking events. To find out more, go to hachettespeakersbureau.com or email HachetteSpeakers@hbgusa.com.

Workman books may be purchased in bulk for business, educational, or promotional use. For information, please contact your local bookseller or the Hachette Book Group Special Markets Department at special.markets@hbgusa.com.

Library of Congress Cataloging-in-Publication Data is available.

ISBN: 978-1-5235-2425-9

First Edition February 2025

Printed in China on responsibly sourced paper.

10 9 8 7 6 5 4 3 2 1

DEDICATION

FROM NATE:

To my dad, Tom Pedersen, for always pushing me (not always successfully) to think more critically.

* * *

FROM LYDIA:

To librarians and educators everywhere: thank you for nurturing curiosity, fostering the sciences, and defending the right to read books. You are our heroes.

AUTHORS' NOTE

Pseudoscience is not intended to be a comprehensive compendium or an exhaustive, evidence-based reference. If it contained every topic under its umbrella, it wouldn't be a book but a library. That said, we acknowledge the existence of deeply harmful pseudoscientific subjects such as racial superiority theory, gay conversion "therapy," Holocaust denial, and many other topics that could not be covered within this short, lighthearted book. Those topics deserve serious discussion and a respectful tone. We are strongly opposed to pseudoscientific ideas deployed to spread hate and suppress marginalized communities. We support and encourage listening open heartedly to organizations fighting back with love, acceptance—and science—such as the NAACP, the American Civil Liberties Union, the Southern Poverty Law Center, the Human Rights Campaign, and The Trevor Project.

Contents

INTRODUCTION vi

PURE NONSCIENCE. 1
Flat Earth 2
Spontaneous Human Combustion. 12
Gasoline Pills 21
Perpetual Motion Machines. 31
World Ice Theory 40
Body Divination 50
Phrenology 64

ALIENS! GHOSTS! BIGFOOT! ATLANTIS! 75
UFOlogy 76
The Bermuda Triangle 90
Crop Circles. 102
Ghosts and Ghost Hunting 112
Cryptozoology 125
2012 Phenomenon 137

WISHFUL THINKING. 149
Cryonics 150
Astrology. 160
Ley Lines. 171
Personality Psychology. 183
Auguries 197
Polygraph 209
Superstitions Hall of Fame. 219

GRIFTERS, NIHILISM, AND DENIALISM 227
Levitation 228
Fake Moon Landings 241
Climate Change Denial 253
Lawsonomy 265
Dowsing 275

SOURCES 285
ACKNOWLEDGMENTS 297
INDEX. 298
PHOTO CREDITS. 308

Introduction

As the world's most recognizable astronomer of the late twentieth century, Carl Sagan was frequently asked the question that nags at many of us on Earth: "Do you believe there's extraterrestrial intelligence?"

Sagan had a well-honed response: "I give the standard arguments—there are a lot of places out there, the molecules of life are everywhere... it would be astonishing to me if there weren't extraterrestrial intelligence, but of course there is as yet no compelling evidence for it."

An intelligent, humble answer. But not a great sound bite.

Inevitably, there would be follow-up questions along the lines of: "What do you really think?"

Sagan would respond, "I just told you what I really think."

The interviewer, still unsatisfied, would press again, "Yes, but what's your gut feeling?"

And then Sagan would shine like the star he was: "But I try not to think with my gut. If I'm serious about understanding the world, thinking with anything besides my brain, as tempting as that might be, is likely to get me into trouble. Really, it's okay to reserve judgment until the evidence is in."

It's a brilliant retort because he's right: The scientific process can be infuriatingly slow. Evidence takes time to accumulate and analyze. New data must be reproducible by other researchers. Scientific consensus is hard-won. And it's very, *very* tempting to jump to conclusions based on gut instincts, on what you want to be true, on what you heard from a friend that might be true, or on what makes the world conform to your particular background and biases. But the scientific process is valuable precisely because it is slow. Because it doesn't leap to conclusions. Because it takes a long time to test hypotheses and build consensus. And that's exactly what makes it so crucial for our understanding of ourselves and the world around us.

· · · · ·

Pseudoscience as defined by Merriam-Webster is "a system of theories, assumptions, and methods erroneously regarded as scientific," and we've believed in it for a very long time. In this book, we take a look back at the long arc of history to observe some of pseudoscience's crowning achievements. After all, it wasn't that long ago when we dragged out chickens to consult before deciding to attack an enemy. Or delayed major political actions because of astrological signs. We've blamed Atlantean space beams for sinking ships in the Bermuda Triangle. The quest to build a perpetual motion machine has been ongoing since the Middle Ages. And we still have a seemingly inexhaustible desire to watch people bring comically designed electronic devices into abandoned houses to prove that ghosts exist. We've even tried to cheat death itself by freezing our corpses. And we've blamed everything—and we do mean everything—on aliens and UFOs.

Aristotle is credited with originating the scientific method in the fourth century BCE. And it's been an uphill battle ever since then. The scientific process continued to refine and develop over the next 2,000 years. It resurfaced in the Renaissance, flowered during the Enlightenment, and exploded during the Industrial Revolution in the late 1800s. Suddenly, the very mysteries of the universe were within our grasp. Refinements in medicine revealed the mysteries of the body. New, exciting discoveries in astronomy and physics began to lift the veil on the world around us. But even as the scientific method took center stage, pseudoscience stuck around, lurking in the back of the auditorium, ready to shout obscenities and throw rotten tomatoes. If anything, it seems to be surging in today's postmodern, post-truth world, where objective truth is no longer in vogue for many. Heck, 25 percent of Americans think the moon landings were faked.

But are we really to blame? Human life is hard. The world can be a strange and terrifying place, replete with uncertainty. We all try to seek explanations for the unusual phenomena that we witness all around us. It's not always easy to trust evidence we can't see with our own eyes. After all, despite all scientific proof to the contrary, let's be honest—the Earth really does seem kind of flat.

May the Science Be with You

We live in a moment awash in conspiracy theories. Powerful people or organizations are routinely blamed for secret plots to influence world events. Earth-shattering secrets are assumed to be kept from the public. Some of these theories are dangerous (think: The Earth is secretly ruled by alien reptiles!). Others may be absurd but are relatively harmless, even fun. (There's a secret chamber behind Mount Rushmore filled with government secrets!)

Aliens. They are responsible for everything.

A 2021 study of conspiracy-inclined personality types conducted by Emory University psychologists Shauna Bowes and Scott Lilienfeld revealed a few shared traits among the conspiracy minded: entitlement, self-centered impulsivity, coldheartedness, elevated levels of depressive moods, and anxiousness. Now consider that roughly 50 percent of the American population believes in at least one thoroughly debunked conspiracy theory, according to the *New York Times*. That's a lot of us. Are we all entitled, coldhearted, and anxious?

The answer, of course, is no. Because there's also a playfulness about some conspiracy theories that allow us to claim our identity, to exert our beliefs in a world where we are constantly being told what to think and do. Staking a claim where you feel like you really know the truth and others don't can feel pretty good. It helps create your identity and, in turn, encourages you to bond with others who are part of the same squad and echo the same belief system. It's a natural inclination, one that stretches back across the millennia to the dawn of time. In a way it makes a lot of sense: When your community is small and vulnerable, you need to watch out for secret plots. And this instinct is still important at times, even though society has grown infinitely larger and more complex. After all, belief in conspiracy theories can have a real-world, tangible benefit sometimes: It can function as a safeguard against real conspiracies. Because sometimes . . . they really do happen.

So, how do you separate out the pseudoscientific wheat from the proven chaff?

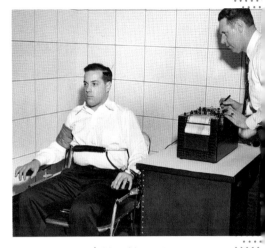

Lie detector test (white shirts not mandatory)

In this book, both harmful and harmless pseudosciences and conspiracy theories are discussed throughout history. The common thread among all of them is that it took time to sort out the truth from the lies, the signal from the noise. As we try to make sense of this chaotic, multilayered mess, remember the key part of the earlier Sagan quote: "Really, it's okay to reserve judgment until the evidence is in."

While we're waiting for that evidence to come in, a healthy dose of skepticism and caution seems logical. It's okay to be uncertain. It's healthy to be a little bit skeptical. As a species, we need to get a lot more comfortable with doubt. We live in a strange and complex world. We should create space for diverse opinions. We should allow consensus to build, a process that takes time. So, we also need patience and more comfort with uncertainty as we wait for the evidence to accumulate.

The next time pseudoscience feels like it's raging all around you, pause for a minute, think with your brain, reserve judgment, and wait for that evidence to come in. If you do that, sorting through the mess is infinitely easier. In this book, however, we've already sorted the mess out for you. In case after case, the scientists have already done the work, the consensus is in, and we're happy to share it (and the proof!) with you.

After reading this book, we're fairly confident you'll keep your egg-laying hens happily in your backyard (versus dragging them out for a consultation before deciding whether or not to attack your neighbor). And that you'll be at least slightly less inclined to connect UFOs, or worse, Atlantis, to any mysterious phenomenon you may encounter in the world. Maybe you'll even enjoy just soaking in the gothic vibes of that creepy abandoned house in the neighborhood and leave the ghost-hunting devices behind.

Don't be surprised if this book inspires you, too. Maybe you want to try your hand at building a perpetual motion machine, knowing full well the unlikely science behind your efforts . . . but hey, who's to stop you from tinkering around in your garage? Everyone needs a hobby. And if that hobby involves prowling the backwoods of Washington State looking for signs of a large humanoid creature previously lost to science, well, at least you'll get some good exercise in the great outdoors. There are certainly worse ways to spend your time.

In the meantime, sit back and let us dispel a few myths with actual science. It just might save you an expensive and disappointing trip to finally find the Loch Ness Monster.

Carl Sagan, brilliant astronomer, who could be happy posing before just about any background.

Pure Nonscience

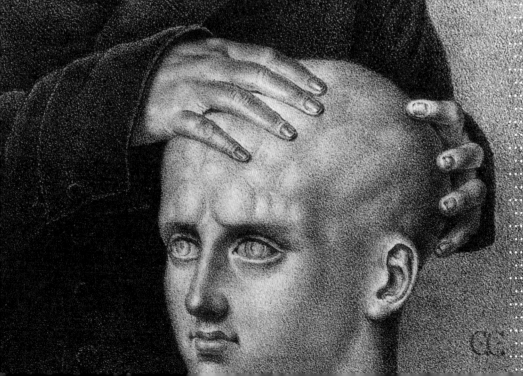

Flat Earth

Mike Hughes led a seemingly quiet life in the arid Apple Valley of Southern California, cohabitating comfortably and simply with four cats he adored. But his résumé revealed a varied, eccentric background in stark contrast to his simple home life. Hughes had been a limousine driver for more than two decades and a NASCAR crew chief before that. He also had a daredevil streak, once jumping his Lincoln Town Car stretch limo for a Guinness World Record of 103 feet.

By 2020, he had added a new line to his unusual bio: self-styled rocketeer bent on proving once and for all that Earth was flat. To accomplish this, he planned to blast himself into the sky via a homemade rocket, thereby providing the truth to the world of skeptics who believed the Earth was globe-shaped.

Hughes openly scoffed at the "scientist" part of the moniker "rocket scientist," claiming in a 2019 documentary called *Rocketman: Mad Mike's*

Mission to Prove the Flat Earth that "Me saying I'm a self-taught rocket scientist is a misnomer because I really don't believe in science. I believe in formulas, some math . . . I just want the raw truth. And I believe in the Flat Earth model."

So "Mad Mike Hughes," as he was often called, set out to experience a firsthand view of the world to confirm for himself whether it was truly flat or spherical. For years and years, his hope was to build his own rocket and send it up to the Kármán line, 62 miles (100 km) above sea level, where the Earth's atmosphere ends and space begins. At that elevation, with supplemental oxygen to breathe and a nice view of Earth, Hughes could finally put to rest the question that most people spent little time considering: Was the Earth really flat?

Hughes's early work involved making a rocket propelled by a superheated water tank that spewed steam on launch. In his first attempt at propelling himself skyward in 2014 in Arizona, Hughes crafted a narrow 15-foot-long rocket with stabilizing tail fins and barely enough room to fit a human inside. It was painted white and decorated with a small number of advertisers and his website—MadMikeHughes.com—in large lettering. The rocket sprang a pin leak at the last minute, but instead of aborting the mission, Hughes jumped into the rocket and cleared the area for the launch, insisting on flying before it blew up. The craft reached 1,374 feet in less than a minute at an estimated 350 mph or more. After the parachute was deployed (and nearly got shredded in the process), Hughes landed hard at 60 mph. The impact injured him so badly he reportedly needed a walker for two months and "his balls were black for a month," according to Waldo Stakes, a fellow amateur rocket enthusiast. The rocket was wrecked.

But! Now that he knew the steam rocket worked, the next steps were . . . faster and higher.

Hughes hoped to build a rocket that would reach a speed of at least 500 mph. After fundraising from

Mad Mike, ready to go toe to toe with Aristotle

fellow flat Earth supporters, he launched himself in another rocket that reached a height of 1,875 feet at a speed of approximately 350 mph. But that wasn't enough. Remember, Hughes's goal was 62 miles, or 327,360 feet. Quite a ways to climb. Before he got that high, there were still more trials to be done—and a lot more money to raise.

Once his next rocket iteration was tested and ready, the plan was for a balloon to first carry it 20 miles above the Earth, where the rocket would fire, propelling it to the Kármán line. Hughes called his theoretical contraption a "rockoon." And from the cockpit of said rockoon, Hughes would finally be able to discern the truth.

In 2018, he remarked, "I expect to see a flat disk up there . . . I don't have an agenda. If it's a round Earth or a ball, I'm going to come down and say, 'Hey guys, I'm bad. It's a ball, okay?'"

The History of Life on a Pancake

Here is a core belief among those who think the Earth is flat: In order to be convinced that the Earth is spherical, they need to see it for themselves. To flat Earthers, it seemingly doesn't matter that a spherical Earth has been a commonly accepted idea since the time of Aristotle (384–322 BCE), who made his observations while noting that the constellations of stars change depending on where the observer is. In the third century, the Greek polymath Eratosthenes accurately calculated the circumference of the Earth, as well as the Earth's axial tilt (it's currently measured as oscillating between 22.1 and 24.5 degrees on a 41,000-year cycle).

Around the tenth century, Persian scholar Abū al-Rayḥān Muḥammad ibn Aḥmad al-Bīrūnī calculated the radius of the Earth through measurements of hills and horizons. Oh, and European explorers circumnavigated the globe via the oceans without ever falling off the edge. (Contrary to popular belief, Christopher Columbus did not think the Earth was flat—though he did think it was far smaller than it is and had hoped to reach Asia by sailing west.) And lest we forget, Earth has been endlessly photographed from space to be revealed as the blue and green marble that it is. And so, repeatedly and scientifically, humans have verified the sphericality of our planet.

The shield of Achilles, or a fancy Greek pizza

Throughout history, many have believed in the flat Earth view of the world. Greek poet Homer described the shield of Achilles as having a disc version of the world displayed. Ancient Norse and Germanic peoples subscribed to a flat Earth cosmography. Ancient Chinese philosophers saw the Earth as flat or slightly rounded and square prior to the European discoveries of the spherical model in the seventeenth century. And sure, okay, as a very tiny organism on a ginormous sphere far larger than yourself, Earth can indeed look flat. It does not feel as if we're all standing on a beach ball.

Our current flat Earth beliefs were partially planted when English writer Samuel Rowbotham published a pamphlet in 1849 entitled *Zetetic Astronomy*

As far as the eye can see . . . and no farther, apparently

("zetetic" meaning "inquiry" or "investigating"). One of his favorite ways to "prove" that the Earth was flat was by showing people the Old Bedford River with its 6-mile-long drainage canal. Because the curve of the Earth means an average person can see only about 3 miles to the horizon, Rowbotham (who wrote under the pseudonym "Parallax") showed them you could see boats at the end of the 6-mile canal. But Rowbotham and others didn't understand that due to air density and bending light, what they were actually seeing was a mirage, or optical illusion. It's the same reason why a setting sun looks distorted as it submerges below the horizon line instead of being a clean, perfect circle.

Englishman Alfred Russel Wallace, a surveyor, answered a challenge in a weekly journal, *Scientific Opinion*, which offered £500 to prove that Parallax was wrong. Setting a sight line higher to reduce the refraction, he added a series of disks on poles along the canal, then showed that the disks halfway along the canal were slightly higher than the ones farther away, proving the curvature of the Earth.

But Wallace's proof didn't really matter. The man who offered the money was a devotee of Parallax named John Hampden, who was outraged at the challenge to his flat Earth worldview and hounded poor Wallace for more than 20 years, libeling him in scientific publications and threatening him as well. Hampden once went as far as sending a letter to Wallace's wife stating, "If your infernal thief of a husband is brought home . . . with every bone in his head smashed to a pulp, you will know the reason." Multiple jail stints didn't stop the harassment. (Generally, flat Earth believers are a harmless bunch, but this historical example certainly gives pause.)

Later, Lady Elizabeth Blount became a member of the Universal Zetetic Society in 1893, which kept the flat Earth conversation and beliefs well fed and lively with journal publications. Her position in society drew many upper-class members. She further enjoyed writing flat Earth verses, ultimately turning some of her poetry into a flat Earth operetta, in which non–flat Earthers are called "Globites." With the assistance of a photographer, she had a picture taken of the Bedford River that would soon become a famous piece of "proof" of the flatness of Earth, given that the distance of the light at the horizon was supposedly too far to see if the Earth was a sphere. The International Flat Earth Research Society was eventually spawned from the Universal Zetetics and launched anew in 1956 in Dover, England, but didn't make particularly large waves. It wasn't until the internet allowed for the easy spread of conversation and questioning that the modern iteration of flat Earthers surged with vigor.

If there's one thing that flat Earth believers agree on . . . it's that they disagree. There are many different views on how a flat Earth works, and much of what follows is hotly debated. But there are nonetheless overlaps in the various flat Earth "truths." Few flat Earthers consider Earth to exist on an infinite plane. Most believe we live on a rounded pancake of land and oceans with the North Pole at the center, surrounded by a wall of ice (Antarctica) that keeps the seas from flowing off into space (and without which the Earth would resemble Marvel Comics's depiction of Thor's home planet of Asgard, with water flowing off into outer space). Some believe the atmosphere is sheltered beneath a hemispherical glass dome, with the sun acting like a spotlight moving around the North Pole. Apparently (and confusingly!), the sun and moon are typically seen as spherical. But the moon is apparently only 31 miles in diameter and about 3,400 miles away (versus the 238,900 miles away we know it to be).

A Barely Inconvenient Truth

But what of the immense amount of photographic proof that the world is clearly not flat? Well, according to the flat Earth crowd, photos of Earth since 1954 and onward were all created through composite photographs, which means they can't be trusted because none are a single photo. No matter that single photos taken from the International Space Station, for example, clearly show the curvature of the Earth from various viewpoints. Take this a step further and consider that, if a true lunar eclipse happens when the Earth's round shadow passes over the moon when it's between the Earth and the sun, how does that work if the Earth is flat? According to flat Earth thinking, apparently a "shadow object" causes all lunar eclipses, not the moon. Many flat Earthers further believe that gravity is not a real thing, choosing instead to subscribe to the idea that Earth is accelerating upward at a rate of 32.17 feet per second, buoyed by a force called the "aetheric wind."

Many flat Earthers also claim to have firsthand knowledge of the Earth's flatness. This is an oft-described reason for global skepticism. Meaning: You're a fool if you believe the lies being fed to you by your addled science teacher and, oh, NASA. Examples of this firsthand knowledge include the following:

Circular arguments, or a rounded view of Earth?

- If you hold a ruler up to the horizon of a large lake, the horizon is flat, not curved. (FYI: The horizon will never look curved below an altitude of 35,000 feet.)

- If Earth is spherical, why don't airplanes constantly have their noses pointed downward so they don't fly off into space? (We are flying over a planet with a diameter of 7,917.5 miles, not a planet the size of Texas.)

- Why can you see a ship on a lake or ocean farther than about 3 miles away, which is the distance that a 6-foot-tall human can see to the horizon before the curvature of Earth impacts their line of vision? (One, because the top of the ship is taller than the horizon; two, due to refraction. Light waves bend over the curvature of the Earth, allowing us to see a mirage of distant objects just beyond the horizon. But there are limits. We can't use a telescope at the top of the Empire State Building to see the Eiffel Tower, for example.)

- If the Earth is a globe, why can't you fly directly from Chile, across Antarctica, to New Zealand? (The shortest distance between Chile and New Zealand is not over Antarctica.)

- Antarctica is supposed to have 24-hour sunlight in the summer, but no one ever goes there to prove this is really happening. (Can't argue with this one. We don't want to go there either!)

The world is chock-full of evidence that the Earth is indeed a globe. For example, the ground is not accelerating upward at 32.17 feet per second (9.8 m/s). If it were, then we would be approaching the speed of light, because remember that "acceleration" means the planet would be going faster and faster upward, not at a constant speed. So, we'd all be dead and flatter than the flat Earth from g-forces if this theory was true. Furthermore, it would not explain why gravity is stronger at the North and South Poles (which is why NASA launches rockets as close to the equator as it reasonably can). Lighthouses are also made extra tall so ships can see them from greater distances before the drop-off due to the Earth's curvature.

And there's further evidence in space. The constellations change depending on what part of the planet you're on. And given modern flight innovations, we could have visited the edge of the Earth-pancake by now and—using a flat Earth tenet—seen it with our own eyes. Space stations, satellites, airplane travel, radio signals, and mountain climbing have all confirmed the spherical world on which we live.

UNITED NATIONS

But these glaring facts are either ignored or argued down with tried-and-true methods. Here are a few: Flat Earthers will offer alternative explanations for each fact proving the Earth is round; they will bring up US government secrecy as a blanket way to explain why you can't trust science agencies like NASA; and unless they can take the photo themselves from the International Space Station, any space photos of Earth aren't real proof. They will additionally say that believing science taught in schools is akin to brainwashing—that we need to decide for ourselves what's real. And so on and so on. Apparently, the rest of us are "globetards" (their word, not ours) for believing these myths.

Let's also not forget that every world government, plus scientific organizations such as NASA and ESA, all science textbooks, and most every human on Earth believes that our world is spherical. How does that translate to the flat Earth crowd? To them, it means that there is a mighty powerful conspiracy hiding the truth and brainwashing the public. But apparently the "truth" creeps out here and there. Consider the United Nations logo, for example, which is a flat disc with all the countries around the arctic circle—and with no Antarctica shown because it's clearly at the edge of the disk. The truth in plain sight! We are all living in a "post-globe reality," evidently.

The recent and growing distrust of authority contributes to the flat Earth cause, of course, and there's a startling resonance with other conspiracy theories, whose believers sometimes overlap with flat Earth believers—Pizzagate, the Sandy Hook school shooting "hoax," etc. And with the near-ubiquity of chat rooms and YouTube videos creating chatter and confirmation bias, it's no surprise that the flat Earther ranks stay well stocked. As with all theories that dabble in conspiracy, it's difficult to battle believers in the realm of logic—because their logic is a moving target. Conspiracy-minded folks cherry-pick what facts they wish to cite or fight, and often set the bar excruciatingly high for the scientific community but then very low for what they consider to be true.

Which invariably takes them back to experiments that show the so-called truth. "Trust only what you can see for yourself" is a common mantra—not the lies fed to you by authority figures, not the millennia of scientific validation. So when in doubt, build your own rocket and do your own observation from 62 miles above Earth.

Which is exactly what Mad Mike Hughes hoped to do. He had another launch planned in 2020, one that he hoped would go higher and faster than his previous attempts, as he continued to perfect his rockoon. His friend Waldo Stakes wondered about his insatiable desire to make the project work, implying that he might be doing it for success, money, and adulation.

"Mike's dad neglected him," Stakes said in the 2019 *Rocketman* documentary, talking of Hughes's background and what drove the daredevil. "He never got any attention as a kid. He, uh, been driving a limousine for so long . . . he wants to be in the back."

Hughes was often broke, estranged from some of his family, in between jobs, and he had cars that got repossessed during these years. For believers in fringe theories, thinking for yourself can sometimes mean isolation—it can be lonely in the echo chamber. Some people worry that flat Earth believers are abandoning family, friends, and regular society for the sake of their beliefs.

For Mike Hughes, living the life of a daredevil believer meant that he selectively chose his science and math and mixed that with a tendency to ignore basic safety measures—like using automatically deploying parachutes in case he was unconscious or otherwise incapacitated during a flight. When his newest rocket took off in February 2020, it scraped the steel ladder that he used to climb into it. One of the parachutes then deployed too early and pushed the rocket off course. Some believe that the impact of the rocket striking the ladder knocked Hughes out. His rocket soared over the California desert and across an impossibly blue sky before smashing into the ground and killing Hughes. The reserve parachute never deployed.

It was an inglorious but perhaps fitting end to a person who once said, "There's something about dreams. It's got to be big enough to scare you . . . And for that kind of dream, it takes everything. It takes all your time, your income, emotionally, physically. It takes everything from you. And it's supposed to. Is it worth it? That's another thing I ponder every day. I don't know. We'll see in the end."

Spontaneous Human Combustion

␣t started innocently enough. Safe at home at her palace in Cesena, then part of the Papal States, the 66-year-old Italian noblewoman Cornelia Zangheri Bandi came down to dinner on March 15, 1731, after an otherwise typical day that she spent in good health and spirits. However, that evening the countess reportedly complained of feeling "dull and heavy," opting to retire early to her bedchamber, where she spent several hours in idle conversation and prayers with her maid. When sleepiness (or perhaps drunkenness) finally overcame her that March evening, the countess dismissed her maid and went to sleep.

The next morning, her maid noticed that the countess seemed to have slept past her usual waking time and went to check on her. Upon opening the door to the countess's bedchambers, the maid encountered a horror scene later vividly reported by Paul Rolli in the *Philosophical Transactions of the Royal Society of London*:

> *Four Feet Distance from the Bed there was a Heap of Ashes, two Legs untouch'd, from the Foot to the Knee, with their Stockings on; between them was the Lady's Head; whose Brains, Half of the Backpart of the Scull, and the whole Chin, were burnt to Ashes; amongst which were found three Fingers blacken'd. All the rest was Ashes, which had this particular Quality, that they left in the Hand, when taken up, a greasy and stinking Moisture . . . The Bed receiv'd no Damage . . . . . . The whole Furniture, as well as the Bed, was spread over with moist and ash colour Soot . . . . . . the Soot was also gone into a neighbouring Kitchen, and hung on the Walls, Moveables, and Utensils of it. From the Pantry a Piece of Bread cover'd with that Soot, and grown black, was given to several Dogs, all which refused to eat it.*

Despite the countess's terrible fiery end, the bed, linens, and furniture seemed to have been unaffected by the blaze. The entire room was covered in a foul-smelling soot (read: carbonized body ash), which had also penetrated the kitchen next door to her bedchambers, where it was absorbed by some bread that even dogs refused to eat. Yet there were no further signs of fire damage, only a mysterious "gluish moisture" on the floor.

One thing, however, was certain: This clearly wasn't your usual house (or palace) fire.

Monsignor Bianchini, the prebendary of Verona (cathedral administrator), was called in to investigate. After dismissing the possibility that the fire was caused by a ball of lightning or other natural phenomenon, Bianchini reluctantly came to the horrific conclusion that the blaze had originated inside the countess's body.

Dickens's Fiery Obsession

No less a prominent citizen of the nineteenth century than Charles Dickens himself was a believer in spontaneous human combustion, reading everything he could on the topic. Dickens was so affected by the cases he read about that the concept ended up in his fiction. In *Bleak House*, his long novel that centers on an extended legal case, the alcoholic rag dealer Mr. Krook meets a fiery end from spontaneous combustion. In the eerie aftermath, two other characters find Krook's ashes in a room where grisly soot still falls from the air, the stench of burnt fat permeates, and a yellow, grisly liquid colors one of the windowsills. Nothing was left of Mr. Krook except for something that resembled a "small charred and broken log of wood." (If the overall scene sounds familiar to you, it's because Dickens was highly influenced by the report of Countess Bandi's death.)

Dickens's reputation for realism, however, came under attack after the scene was first published in serial form in December 1852. The critic George Lewes, Dickens's friend and George Eliot's lover, took pen to paper in an epic Dickens takedown published in the *Leader*: "Dickens, therefore, in employing Spontaneous Combustion as a part of his machinery, has committed this fault of raising the incredulity of his readers; because even supposing Clairvoyance and Spontaneous Combustion to be scientific

> "The fire was caused in the Entrails of the Body by inflamed Effluvia of her Blood, by Juices and Fermentations in the Stomach, by the many combustible Matters which are abundant in living Bodies, for the Uses of Life; and finally by the fiery Evaporations which exhale from the Settlings of Spirit of Wine, Brandies, and other hot Liquors in the Tunica villosa [inner lining] of the Stomach, and other adipose or fat Membranes."

In other words, the Countess had just entered into the historical record as the first documented case of spontaneous human combustion.

truths, and not the errors of imperfect science, still the simple fact that they belong to the extremely questionable opinions held by a very small minority, is enough to render their introduction into Fiction a mistake."

Dickens leapt to his own defense, adding a preface to the next serial part of *Bleak House* pointing out the ample scientific speculation about the possibility of spontaneous human combustion. He made reference to the Bandi case, among others. A Victorian war of letters followed, with neither writer retreating until a year later when Dickens, in the final edition of the novel, published the last word on the subject: "I have no need to observe that I do not willfully or negligently mislead my readers and that before I wrote that description I took pains to investigate the subject . . . I shall not abandon the facts until there shall have been a considerable spontaneous combustion of the testimony on which human occurrences are usually received."

Mr. Krook's fiery demise

Or had she? As with so many mysterious deaths attributed to spontaneous human combustion over the years, there were other forces at work. The countess was well-known for her love of brandy and almost certainly drank freely in her long wind-down from the "dull and heavy" feeling that overcame her at dinner until her bedtime some three hours later. But the countess's love of brandy extended beyond simple consumption. As the prebendary reported: "The old lady was used, when she felt herself indisposed, to bathe all her body with camphorated spirit of wine . . ."

Like others in the eighteenth century, the countess believed that camphorated spirits were an effective skin tonic. In those days, camphor was

Camphorated oil: good for soothing your skin . . . or setting it on fire

made from the distillation of the bark and wood of the camphor tree. Distillation produced the chemical camphor in solid form, which was known for its strong taste and smell. Camphor was then mixed with alcohol to create "camphorated spirits," which were sold for medicinal use as a skin tonic. (Not, by the way, as ineffective as you might suppose—camphor does have some anti-inflammatory and anti-itch properties.) Camphor mixed with alcohol, however, also has a distinct disadvantage: It's *highly* flammable.

You can probably guess where this is heading. The countess followed her usual bedtime routine of covering her body with camphorated spirits, probably while also having drunk more than her share of brandy. And then she went to bed . . . in a room lit entirely by oil lamps and candles. Sometime during the night, the countess likely arose to use the bathroom, or open a window. In her sleepiness, she possibly bumped into her own oil lamp and spilled the oil on herself, a conjecture hinted at by the location of the empty oil lamp near her remains. From there, the tragedy quickly unfolded. The fire from the oil lamp would've ignited the oil and camphorated spirits on her body, quickly consuming the poor woman in an aggressive burst of flames. In a quick and furious ending, the countess would have burned to a crisp. And so a case of spontaneous human combustion was likely a case of a well-honed skin care regime gone horribly awry.

Going Up in Flames

Spontaneous human combustion is the pseudoscientific belief that human beings can, at random, burst into flames without an obvious cause of ignition. The thinking goes that some mysterious force at work inside the body can set it on fire from the inside out. The notion caught the public imagination in the early nineteenth century, when, sadly, death by fire was still relatively common. In the 1800s, lighting and warmth were generated by flames, most structures were built of wood, and most clothes were flammable. Early nineteenth-century US and Great Britain were basically a pair of tinderboxes.

Layer in a few mysterious, fire-based deaths and you've got fertile ground for the imagination to seek explanations outside the ordinary. The popular press got wind of "spontaneous human combustion," reported on it breathlessly and repeatedly, and voilà, a growing popular fascination with this grisly means of death.

All this Victorian interest in spontaneous human combustion was connected to a scientific discovery made mere decades earlier: oxygen. After isolating oxygen for the first time in 1774, scientists soon realized that this essential gas was critical for both breathing . . . and burning. This gave rise to the theory that breathing itself was a form of combustion, a sort of continual burning of oxygen taking place in our lungs.

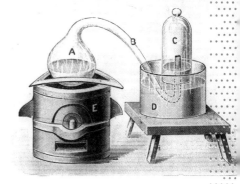

Joseph Priestley's device for isolating oxygen

Despite being entirely incorrect, it's kind of a beautiful theory, no? All these little fires burning away in our lungs? But what's actually happening is that every time you breathe in, you absorb oxygen from the air around you. The oxygen travels down your bronchial tubes to the alveoli, or air sacs, in your lungs, where the oxygen passes through thin walls into the surrounding capillaries, to be picked up by your red blood cells. Perhaps not quite as beautiful as little fires in your lungs, but a lot more practical. For those who believed in the fires-in-your-lungs theory, it wasn't much of a leap to suggest that these slow fires could, on occasion, flare up and overtake their host, perhaps in moments of "heated" anger or "fiery" rage.

Alcoholics were viewed as being particularly susceptible to the phenomenon. Their very organs were thought to be dripping with flammable alcohol. Furthermore, the alcohol they consumed was thought to be transmuted into a highly flammable gas within their bodies, just sort of floating around waiting for a source of ignition.

If you are a drinker, don't worry—your organs aren't dripping with alcohol. But frequent or excessive alcohol consumption certainly puts a heavy strain on them. Nor is there a ball of alcoholic gas floating around your body. Instead, your liver enzymes painstakingly break down alcohol into carbon dioxide and water. Before that process was understood, however, it was little wonder that most of the 50 or so cases of spontaneous human combustion reported in the Victorian press centered on confirmed or suspected alcoholics.

In 1800, the French physician Pierre-Aimé Lair published a scholarly review of cases of spontaneous human combustion, all of which he attributed to alcohol. Lair observed some similarities among the cases he reviewed. In addition to being alcoholics, the victims tended to be overweight and elderly. He also noticed that the combustions tended to be short and rapid in duration. And that the fires often failed to ignite any of their surroundings, leaving behind in their wake a residue (and stench) of burned organic material. Lair further identified those liquors most likely to ignite you, a helpful guide for readers hoping to avoid such a grisly fate. They were, in order: gin, brandy, whisky, and rum.

Sobering Thoughts

The growing popular belief in the connection between drinking alcohol and bursting into flames was not lost on the budding temperance movement. Spreading the fear of spontaneous combustion became another weapon in their arsenal of ideas to stem the tide of nineteenth-century drunkenness. (And, to be fair, they had their work cut out for them. Americans in the nineteenth century drank at roughly three times the rate of Americans today. On average, approximately 1.7 bottles of standard 80 proof liquor [40 percent alcohol] were drunk per person, *per week*.)

Temperance literature became filled with cautionary tales, including that of a Bohemian peasant who died when a column of "ignited inflammable air" issued forth from his mouth, or an Italian man found alone in his room covered in a blue flame, or the Canadian man found roasted alive and who survived a further two weeks "suffering the torments of hell" all the while knowing "he was about to enter its dismal caverns."

All heady consequences to consider before pouring yourself another drink.

The temperance movement almost single-handedly kept alive the concept of spontaneous human combustion well into the twentieth century, long after scientific discoveries had rendered it a moot proposition. The human body is composed mostly of water—approximately 60 percent—and its only really flammable components are hair, methane gas, and fat tissue. All that water, which you might think would help put out a fire once started, actually boils out ahead of the advancing flames. That body fat, meanwhile, is highly flammable and contributes to the "wick effect."

As for the wick effect, get ready. It's pretty gross. Basically, once you catch fire, your body fat acts quite literally as fuel for the flame. And the clothing you wear doesn't help much. Picture an inside-out candle, with the fuel source (your body fat) on the inside and the wick (your clothing) on the outside. As your melting fat continues to drip onto your clothes, the hydrocarbons of the fat continue to add fuel to the flame. And it's just a negative, self-perpetuating spiral from there.

The temperance movement meant business.

The wick effect is locally contained by the continual supply of a highly flammable fuel source, meaning the flames typically don't spread to your surroundings. This is why you see such unusual conditions with supposed cases of spontaneous human combustion, where the body is burned to an unrecognizable crisp, but the couch on which they were sitting is only minimally damaged. Eventually, just as a wax candle will eventually burn itself out, so will a human one.

Most scientists argue that deaths attributed to spontaneous human combustion are actually examples of the wick effect caused by an undetected source of ignition. It's not an accident, they argue, that many victims of spontaneous human combustion are found close to a fire source, suggesting they lit themselves on fire—while trying to start a different fire.

Take the case of Mary Reeser, whose ashes confounded investigators when they were discovered in her Florida apartment in 1951—just her skull, backbone, and part of her left foot still recognizable. Mary, who both smoked cigarettes and used sleeping pills, most likely took a pill, lit a cigarette, and fell asleep, at which point the lit cigarette dropped onto her clothes and ignited the wick effect.

Or the case of Henry Thomas, a 73-year-old Welshman whose ashes were found next to his skull along with portions of each leg in his council flat in Ebbw Vale in 1980. The wick effect again.

Or Michael Faherty, the Irishman found burned to death near his fireplace in Galway in 2010. There was so little left of Faherty's body that the coroner could find no plausible explanation for the death and filled in "spontaneous human combustion" in the official report. Once again, the most likely culprit here was the wick effect, with the source of ignition (a match to light the fire?) having been lost to the flames.

· · · · ·

Human combustion is never "spontaneous." It is, however, always tragic (see above). As such, it should be treated with respect. It's not caused by alcoholism, nor by other pseudoscientific theories, such as poltergeists, or, as suggested by the 1995 book *Ablaze* by Larry E. Arnold, initiated by a heretofore unknown subatomic particle Arnold called "pyrotron."

The horrific deaths attributed to spontaneous human combustion do have some sad commonalities: They typically occur among people who are older, who are living alone, and who don't have anyone nearby to help. So perhaps the real lesson here is one of charity—an important reminder to check on your neighbors, particularly those living alone, and to watch out for the vulnerable people in your community. In a society with more regular and obvious means of support and connection, it isn't just the grisly ends attributed to spontaneous human combustion that can be prevented, but a variety of other byproducts of loneliness and isolation as well.

Gasoline Pills

asoline prices. Like the weather, they're a topic of perennial interest to humankind—and as good a conversation opener with just about any stranger. We've been grumbling about gas prices for more than a century—since 1908, to be precise—when the first Ford Model T rolled off the production line. In 2022, the *New York Times* reported that our national mood in the US even rises and falls in sync with gas prices.

With more than a century of griping about the price of gas, it's little wonder that the first effort to "solve" the gas problem happened in April 1916, eight years after cars with internal combustion engines arrived on the scene. A German immigrant to the US named Louis Enricht attracted national headlines when he claimed to have developed a water-based substitute for gasoline. Gas prices at the time were $0.21 per gallon (roughly $6.05 in 2023 dollars). This was considered so high that anyone who could develop a cheaper alternative would reap a fortune and be declared—then as now—a national hero.

That April morning in 1916, Enricht invited several New York City–based reporters out to his house on Long Island, where he planned to show them something truly special. Gasoline, Enricht claimed, was about to be a thing of the past. Through his sheer brilliance and inventiveness, he had discovered a way to transform water into fuel. The reporters laughed, but having traveled all the way from the city to Long Island, they would at least take the time to watch this bold claim fall flat on its face.

Enricht gave his introductory speech next to a small European-made car he had at the ready in his driveway. He invited the reporters to confirm the car's gasoline tank was, in fact, empty. It was. He invited the reporters to inspect the car for any signs of a hidden gasoline tank. They found none. He then proceeded to fill a nearby bucket with water. Next, he revealed a small bottle of green liquid. This was Enricht's magic formula, the "gasoline pill" that would transform a modest bucket of water into a spectacular source of fuel. He emptied the green liquid into the water, sloshed it around, and then poured the mixture into the car's fuel tank.

It smelled oddly like cyanide. Tasted like it, too. One of the reporters gamely tasted the mixture and commented that its flavor was like "bitter almonds," a dead giveaway to the presence of cyanide for anyone who has read a few Agatha Christie novels. Enricht confirmed the presence of the poison, which he said was a purposeful addition to mask the smell of the mysterious ingredient that made this transformation possible. (Any comment he may have offered on how he had just knowingly let a reporter sip cyanide was lost to history.)

Enricht then hopped into the car and turned the ignition over. The vehicle came sputtering to life, leaving the reporters surprised, even astonished. Perhaps Enricht really *was* onto something?

He then invited the reporters to take the car out for a drive, which they did. After puttering around the streets of Farmingdale, Long Island, for an hour, the reporters returned to Enricht's house as converts. The next day, the newspapers were full of breathless accounts of this bold new discovery that could transform water into fuel.

And just like that, Enricht was a star.

From Water into . . . What?

In the mold of a true salesman, Enricht refused to provide answers to the countless inquiries he received in the following days as more reporters, automotive industry representatives, and curious members of the public inundated him with questions. His secret, he claimed, was in the formula he had invented. Even in those early days of the automobile, the car industry was already a powerful economic player, and innovators in the growing field were poised to make a fortune. Fearing that someone would try to steal his secret formula, Enricht carried around a pistol for self-protection.

"The secret is mine and until somebody gives me a fair reward and promises to make [it] a public benefit, I'll guard my rights to it even if I have to use this," he said, in reference to his gun.

Enricht's stalling tactics worked. Interest in his formula grew to such a fever pitch that it wasn't long before Henry Ford himself turned up on Long Island for an in-person demonstration. Suspicious initially, Ford inspected the automobile closely for signs of a hidden tank, paying close attention as Enricht added the mysterious green liquid to a gallon of water before pouring the mixture into the gas tank. Sure enough, the ignition turned over and the car was off and running.

Amazed by what he saw, Ford offered to buy the secret formula. Enricht demurred, declaring that he was not interested in selling the rights until he had secured a patent. Thoroughly convinced of the possibilities of this new

Henry Ford allowing an unidentified hitchhiker to practice driving a stick shift

formula, Ford handed Enricht a check for $10,000 to help secure his interest. When the time was right and Enricht had obtained his patent, he could cash the check and sell the formula to Ford.

For his part, Ford intended to release the formula to the public. From altruism? Perhaps. More likely, Ford saw the formula as a way to get more people to buy his cars. If gas was essentially free ($0.01 per gallon was the estimated cost of Enricht's water-based formula), then it would remove one more barrier from legions of would-be buyers of Ford automobiles.

"If Mr. Ford carries out his intentions, as he has said he will, everything will be all right," Enricht said. "I will tell you this much: Mr. Ford's motives in this whole matter are entirely unmercenary. If he were to buy my formula tomorrow it would be given out . . . to all the people, and I'm with him on that."

Benjamin Yoakum, who contacted the president and hired the Pinkerton Agency . . . before just opening up an envelope

But other suitors came calling shortly thereafter—and the checks got bigger and bigger. One investment banker named Benjamin Yoakum offered Enricht $100,000 but insisted on being able to see the secret formula. Enricht was still hesitant to reveal it until he'd secured a patent, so a compromise was proposed: Enricht would give Yoakum the secret formula in a sealed envelope under strict instructions not to open it until the patent was obtained.

With the deal struck, Yoakum reached out to his old friend President Woodrow Wilson to tell him that he'd secured the future of the automobile industry and that gasoline-free cars were about to become a reality. Wilson was reportedly delighted with the news.

But Enricht continued to drag his heels in his supposed efforts to obtain a patent. Excuses were offered to Yoakum, but without any concrete evidence of steps being taken to move the process forward. Yoakum grew increasingly suspicious of his new business partner, eventually hiring the Pinkerton detective agency to investigate Enricht. What they found was less than reassuring.

Fueling a Fraud

It turned out Enricht had a suspicious past. While living in Chicago in 1903, he'd been implicated in a land swindle scheme where he exchanged worthless land deeds for $500 a pop. The scheme netted $50,000 before the authorities shut it down. What's more, Enricht was found to be consorting with a German military attaché in the midst of World War I while Americans were fighting the Germans overseas. The detectives thought it a distinct possibility that Enricht was a spy.

After reading the Pinkerton report, Yoakum tore open the envelope with the secret formula inside. To his dismay, but perhaps not surprise, he found that it contained only some blank pieces of paper. Enraged, Yoakum vowed to prosecute Enricht for fraud but unfortunately died before Enricht could be brought to trial.

With Yoakum dead, Enricht was never forced to reveal his formula, and he continued insisting he'd found a magic gasoline pill. His reputation damaged but not entirely ruined, Enricht amazingly continued in business. In 1920, he pulled off yet another hoax, convincing a round of investors that he could turn peat into gas just as he did with water. He was less careful this time, however, and a close observer found a hidden fuel line running to

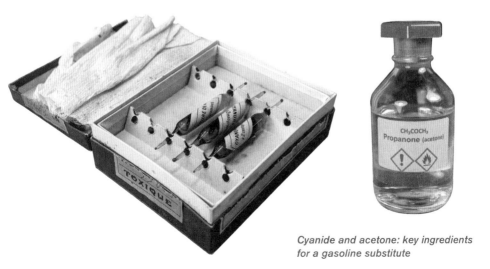

Cyanide and acetone: key ingredients for a gasoline substitute

the tank. By 1922, Enricht was finally convicted of fraud and served several years in prison. He was released shortly before he died, at age 79, without ever revealing the true nature of his secret formula.

Scientists have since figured it out. Enricht stumbled onto a discovery also made earlier by Thomas Edison: that a mixture of acetone, liquid acetylene, and water would, in fact, power an internal combustion engine. But acetone, commonly used as a nail polish remover, has a distinctive and recognizable smell. So Enricht added cyanide to mask the scent of the acetone, which might otherwise have given away his secret formula. While this mixture does indeed work in short bursts, as you can imagine, continually running water through a car engine doesn't exactly result in long-lasting performance. The other problem was that, far from being free, acetone is actually quite an expensive ingredient.

The dreams of penny-a-gallon gasoline were not to be. Or at least not yet.

The Backyard Bandit

About 30 years after Enricht, another man attempted to meet the challenge of transforming water into gasoline. Guido Franch, a coal miner from Illinois, revealed to the press in the early 1950s that he, too, had developed a secret formula to achieve this feat. Franch called his formula "mota gas," mota being "atom" spelled backward. (If you're not particularly impressed, remember that this was the so-called Atomic Age, so all things atomic were new and exciting.)

If you stopped by Franch's Livingston, Illinois, backyard around 1955 with some cash to blow as a potential investor, Franch would happily demonstrate his secret formula for you. He'd do so by dumping some sort of mysterious powder into water, turning the water green—that's how you knew it would work—then the liquid would be emptied into the fuel tank of a lawn mower, which would roar to life with the mysterious liquid running through its veins. Hopefully, this demonstration was enough to get you to part ways with a bit of cash.

Franch was always on the lookout for potential investors, but also nervous about attracting the attention of Big Auto. He was sure they would

A lawn mower not being powered by mota gas

steal his formula and keep it from the public... and not pay him a cent.

Part of Franch's success as a fraudster was that he had a great origin story for Mota fuel. He claimed he hadn't developed it himself, but instead had worked part-time in the laboratory of a German scientist, Dr. Alexander Kraft, who had emigrated to the United States between the world wars and settled in Franch's hometown of Livingston, Illinois. It was this mysterious Dr. Kraft who had discovered a secret formula, based on coal, that could transform water into high-octane gasoline. Allegedly, Kraft shared this secret with Franch shortly before he died, and it was now Franch's mission to reveal this secret to the world. But of course it would help a lot if you would invest $1,000 in Franch's fledgling company . . . in exchange for 1 percent of the profits.

Franch did a lot of backyard demonstrations, but not much work on building, or even starting, a company. He attracted some press attention in tabloids and duped a lot of people out of pocket change that he never got around to paying back. Which people didn't particularly like, when they got nothing in return. So eventually Franch was hauled into court—twice, in 1954—where he managed an acquittal, and then again in 1979, when fraud charges brought against him finally stuck.

The mysterious green powder? It was actually vegetable dye . . . and the water was actually aviation fuel, substituted in a clever sleight-of-hand trick. As for the $20,000 Franch had gathered from gullible investors? Long gone.

Franch was ultimately convicted and sentenced to five years' probation.

Still Not Cooking with Gas

The idea of virtually free gasoline is so appealing that it resurfaces every few decades. The most recent example was the Australian company Firepower International, which, under the management of Tim Johnston, had a surprisingly good run of it. From 2004 to 2007, when it finally went belly-up, Johnston was able to convince huge swaths of the public that his company had developed a fuel tablet, or gasoline pill, that could cut the emissions of automobiles and increase the longevity of fuel. Firepower International, registered in the Virgin Islands, but actually run from an industrial complex in Perth, was able to attract more than $100 million in investment seed money and a range of high-profile endorsements. Firepower's benefactors included the likes of actor Russell Crowe, who in 2006 declared on *The Tonight Show with Jay Leno* that Firepower was sponsoring the Australian rugby team he owned . . . to the tune of $3 million over three years.

The company finally collapsed in 2007 in a flurry of drama and international news headlines. No assets of the company could be retrieved. Johnston himself declared bankruptcy and was banned by the Australian government for 20 years from managing companies. And no efficacy of the company's magic gasoline pill was ever demonstrated.

No Gas, All Brakes

The reason no gasoline pill works is because, scientifically, they're basically an impossibility. Gasoline is a hydrocarbon fuel made up of, yes, hydrogen and carbon molecules. Water consists of hydrogen and oxygen molecules. Much like oil and water, gasoline and water don't mix. Nor is there anything you could dissolve into water that would turn it into a combustible fuel. Unless you were to break the actual water molecule up into burnable oxygen and hydrogen gas. Though you'd need to harness nuclear reactions as powerful as those that take place inside of stars (like the sun in our solar system) to transmute the hydrogen into carbon. That's a lot of trouble for a gallon of gasoline. But just adding a pill's worth of anything

The Ford Nucleon: not featured in the next Fast and Furious *movie . . . but it should be*

to a tank of water? The water itself would ultimately cause the engine to sputter and stall.

The idea of a nuclear-powered car, however, was once seriously considered. In 1957, Ford debuted plans for the Ford Nucleon, a sleek, aerodynamic car that would be powered by a small nuclear reactor. This was during the height of the Atomic Age, years before the Chernobyl disaster, when limitless nuclear energy seemed a safe, exciting, and inevitable part of the future. The Ford company believed that, over time, nuclear fission technology would become more affordable and could be used on a smaller scale. So small, in fact, that the company predicted that in just a few more years they'd be able to fit a nuclear reactor in the trunk of a car. And when that happened, gasoline, they were sure, would become obsolete.

The Nucleon held a great deal of promise: It would be quiet, emit no harmful vapors, and offer incredible fuel mileage—much greater than the most efficient cars then on the market. The car's nuclear reactor would work similarly to a nuclear submarine's, just on a smaller scale. It would use uranium fission to heat a steam generator, which would in turn convert stored water into highly pressurized steam to drive a set of turbines. The turbines would then provide the torque necessary to propel the car into motion and also power an electrical generator to provide energy to the car. Through condensation, the steam would be transformed back into water in a cooling cycle, and then it could be reused again in the steam generator. It was a near-perfect closed loop system, allowing the reactor to continue to produce power for as long as the fissile material (the uranium) remained.

Ford estimated that this system would last for about 5,000 miles. So, instead of your usual 5,000-mile oil change, you'd pull into a nuclear reactor station to have your reactor swapped out. What's more, you'd have your choice of reactor. Concerned about the environment? You could opt for a reactor designed to max out your fuel efficiency. Want to street race your friends? You could instead go for the high-performance reactor and enjoy a mega boost to your speed.

It heralded a future too good to be true. Literally. Nuclear technology would eventually betray its early promise. For the Nucleon to work, its nuclear reactors would have to become very small, an innovation that never happened. An alternative to the heavy lead-based shielding that protects people from radiation would need to be developed—a discovery we're still waiting on today. Large nuclear reactors with heavy lead shielding don't make for a practical car, so plans for the Nucleon were scrapped.

And it's a good thing, too. Can you imagine a world where every minor car accident becomes a radioactive disaster site?

· · · · ·

Oil is an exhaustible resource. We've always known that the age of the gas-powered vehicle was going to be term limited. The future of cars is finally here, and it's electric. Zero-emission target dates have been set in place in many states around the US, some as early as 2050, when many of the cars in the US will already be electric. Such a future contains numerous benefits, including reducing air and noise pollution and lessening the impacts of our increasingly warming climate.

But one benefit that is almost never the focus of the conversation? The disappearance of gasoline pills and the cons who hawk them, exploiting others for profit. These false "solutions" to our gas woes will also soon be a thing of the past.

Perpetual Motion Machines

In 1812, Charles Redheffer abruptly appeared in Philadelphia, ready to make his entrance into the history books. The inventor claimed to have created a perpetual motion machine, one that could run infinitely without an external power source. If he had truly invented such a machine, he would have fulfilled the dreams of mechanical engineers since time immemorial. Redheffer set up shop on the outskirts of Philadelphia, constructing a working model of his machine in a house on the banks of the Schuylkill River. Once the machine was operational, Redheffer was ready to show the curious public . . . all for the cost of $5 a head. A manly head, that is. Redheffer allowed women to view the perpetual motion machine for only $1.

The machine was a hit with Philadelphians, who took to debating the efficacy of the machine in print and parlor rooms around the city. Strong feelings were expressed. Large wagers were made. Charles Gobert, a civil engineer, made an announcement in the *Philadelphia Gazette* on July 12, 1813:

> *I hereby offer, on demand, any bet or bets from $6,000 to $100,000, to the end of proving, in a few days, both by mathematical data and three several experiments, to the satisfaction of enlightened judges, chosen by my very opponents out of the most respectable gentlemen of this city, or of New York, that Mr. Redheffer's discovery is genuine and that it is incontestably a perpetual self-moving principle . . .*

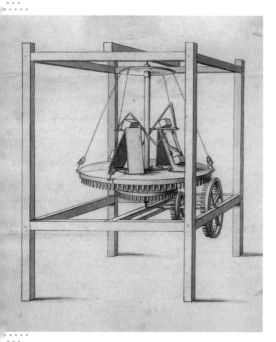

Charles Redheffer's perpetual motion machine

Redheffer's machine was operated by a gravity-driven pendulum, in which the output gear drove a vertical shaft in a whir of seemingly inexhaustible motion. The principle behind it was that an inclined plane could receive continuous downward force from gravity and in turn produce a corresponding horizontal force that could operate some other machine. In essence, it was a "generator" that would power something else. And by all accounts it appeared to work, which caused a minor sensation in nineteenth-century Philadelphia. People queued up to see the machine in action, making Redheffer a tidy little fortune in the process, even with his discounted admission fee for women.

Emboldened by his success, Redheffer applied for government funds from the City of Philadelphia to build and exhibit a larger version of his perpetual motion machine. Not in the habit of simply writing checks for would-be inventors dabbling in perpetual motion, the city insisted first on an official inspection of the machine. On January 21, 1813, eight engineers set off to see if Redheffer's machine truly worked as suggested.

But Redheffer—not at all suspiciously—would allow the inspectors to look at the machine only through a barred window. "They might damage it," was Redheffer's official reason. That didn't prevent the young mechanical genius Coleman Sellers, son of the inspector Nathan Sellers, from noticing a subtle but crucial detail: The cogs on the gears were worn on the wrong side. Meaning: The machine was set up in reverse. While Redheffer claimed his machine was built to power a separate device through a series of gears and weights, the machine itself was actually receiving power from another device.

Bring-your-kid-to-work day paid off for the Sellerses. Rather than call out Redheffer publicly in front of the press, the elder Sellers instead hired a local engineer named Isaiah Lukens to go to the trouble of building a perpetual motion machine modeled on Redheffer's. But with a hidden clockwork motor to power it. Sellers and Lukens then demonstrated the machine to Redheffer, who was convinced it was a genuine perpetual motion machine and offered to buy it. For a lot of money.

The reveal that the machine was operated by hidden clockwork must have been an epic "gotcha!" moment for Sellers, who clearly believed in delivering revenge in icy fashion. Upon realizing that his scheme was unmasked and his time in Philadelphia was over, Redheffer departed the city as abruptly as he arrived. Taking advantage of how slowly news traveled in the early nineteenth century, Redheffer simply reappeared in New York City shortly thereafter, where he again displayed the machine to the public . . . and again began to make a lot of money.

With lessons learned from his experience in Philadelphia, Redheffer made a few tweaks to the machine so the ruse couldn't be as easily detected. What he didn't count on, however, was a visit from the bad boy of nineteenth-century mechanical engineering: Robert Fulton, an internationally renowned engineer and inventor who designed the world's first

Robert Fulton, bad boy of the engineering world, and also very hard to fool

steamboat, submarine, and torpedo. Mic drop. Fulton wasn't about to be taken in by some fraudster. When he dropped by Redheffer's New York exhibition, he noticed a curious detail of the machine—it was oddly unsteady in its movements. It wobbled. Which suggested that the machine was being driven manually, and somewhat irregularly, by a hidden person operating a crank.

Unlike Sellers in Philadelphia, Fulton immediately challenged Redheffer over his machine. Fulton declared the contraption a fraud and said that if Redheffer would allow him to try to expose the secret power source, Fulton would reimburse him for any damage he caused. (Bizarrely, even though Redheffer knew he had a fake machine in operation, he agreed to Fulton's challenge.) Fulton then promptly tore open some boards from the wall behind the machine, exposing a hidden catgut cord that ran to the second floor of the building. Fulton then ran upstairs to determine the origin of the cord . . . where he startled a bearded old man distractingly churning away on a hand-crank with one hand, while trying to scarf down some bread with his other hand. (Because you need a continual supply of carbs to do all that manual labor.) The difficulty inherent in trying to eat bread with one hand while keeping a crank moving at a steady pace with your other hand explained the uneven movements of the machine.

Fulton then loudly and triumphantly declared the machine a fraud. This unleashed the fury of the crowd, who sought their own justice by destroying the trickster perpetual motion machine. Redheffer fled New York City, this time disappearing into the great maw of the nineteenth century, never to be heard from again.

While Redheffer's original machine was ripped and torn apart, the machine that the Philadelphia engineer Isaiah Lukens built based on his model can still be seen at the Franklin Institute in Philadelphia.

Something for Nothing

The dream of perpetual motion persists to this day because it epitomizes an urge we all have: to get something for nothing. Admittedly, the implications are intoxicating. Free energy for all! Goodbye global warming! But the problem with perpetual motion is that it also happens to fly in the face of physics, violating both the first and the second laws of thermodynamics.

The first law states that energy cannot be created or destroyed, only transformed from one form to another. If you are not adding energy to a system, you cannot take energy out and also expect it to operate indefinitely. The second law of thermodynamics related to entropy states that an isolated system will move toward a state of disorder. Entropy, or the idea that not all energy can be harnessed, always increases over time. The more energy that is transformed, the more energy that's wasted. Some energy is always lost (through friction, for example) or dissipates in other ways, so not all of the energy produced can be harnessed. A perpetual motion machine would require energy that is never wasted and never moves toward entropy.

Which is impossible, according to physics. Or, to put a finer point on it, impossible above the atomic level.

What *is* possible is a machine that will operate for a really, *really* long time. But a really, really long time is still not forever. For example, machines operated by ocean currents would appear to be inexhaustible, and for all practical purposes, are inexhaustible within our human conception of time. However, eventually the sun will burn out and when it does, about five billion years from now, all the energy generated by ocean currents, which is ultimately derived from the sun, will also dissipate.

Norman Rockwell's iconic 1920 cover for Popular Science shows an amateur engineer puzzling over a perpetual motion machine.

Perpetual Motion Atoms

In 2016, scientists discovered a new form of matter they called "time crystals." Time crystals are quantum systems of particles whose component atoms are in continual repetitive motion, thus defying the laws of thermodynamics. The system cannot lose energy to the environment because it is already in a quantum ground state. If you want to get technical about it, and, really, who doesn't, the time crystals have "motion without energy," rather than perpetual motion, because the motion of the particles does not represent kinetic energy. That means energy can't be extracted. But really, who cares when there is something as cool as time crystals actually in existence? Cue the 1980s fantasy movie music.

Despite some long-established laws of physics pointing to the contrary, many inventors over the centuries have tried their hands at building a perpetual motion machine. Machines with "overbalanced wheels, rolling weights, water wheels endlessly pumping their own water, inclined planes (such as Redheffer's machine), squirrel cages with steel squirrels forever pursuing a magnet, even rings of balloons inflating under water, hopefully ad infinitum, to lift themselves to the surface," wrote Clifford B. Hicks in *Scientific American*.

Over the years, perpetual motion machine inventors seem to have established an unspoken set of rules. A perpetual motion machine needs to work without assistance except from gravity, buoyancy, or magnetism. (This ruled out using daily variations in temperature or air pressure, or the constant motion of ocean currents.) Ideally, the machine would do something useful, but if not, that was okay, too—it just had to be like the Energizer Bunny—it had to keep going... and going... and going.

A classification system was developed to differentiate the types of machines from each other and identify which law of thermodynamics they sought to violate. A perpetual motion machine of the first kind could produce work without the input of energy, thus violating the first law of thermodynamics. A perpetual motion machine of the second kind could spontaneously convert thermal energy into mechanical work, thus violating the second law of thermodynamics by converting heat into useful work without any side effects. A perpetual motion machine of the third kind is not a machine gifted from extraterrestrials, but instead is a machine that completely eliminates friction and maintains perpetual motion due to mass inertia. This again violates the second law of thermodynamics because such a machine would eliminate any kind of dissipation of energy, which is impossible above the atomic level.

The Never-Ending Quest

The idea of perpetual motion has haunted inventors since the dawn of invention. In the twelfth century, Indian mathematician and astronomer Bhāskara II described a wheel, so-dubbed "Bhāskara's wheel," that he claimed would run forever. The wheel consisted of curved spokes partially filled with mercury. Once the wheel started spinning, the mercury would flow from one side of the spoke to another, forcing the wheel to continue in perpetual motion. (While appealing in theory, it doesn't actually work.)

None other than Leonardo da Vinci himself dabbled with perpetual motion. Some drawings exist of devices that he seemingly thought capable of generating free energy. Though Leonardo was against perpetual motion in theory—"Oh ye seekers after perpetual motion, how many vain chimeras have you pursued? Go and take your place with the alchemists"—he couldn't resist the allure himself, drawing overbalanced wheels and examining those produced by others.

Bhāskara's wheel

Steampunk perpetual motion machine

The earliest British patent for a perpetual motion machine was granted in 1635. But the obsession with perpetual motion machines didn't really take off until the second half of the nineteenth century, when an astonishing 600 patent applications for perpetual motion machines were made to the British Patent Office between 1855 and 1903.

The obsession, fueled by the heady possibilities of the Industrial Revolution, was not limited to the British side of the Atlantic. American inventors were also entranced by the idea. Even the cover of the October 1920 issue of *Popular Science* magazine depicts a Norman Rockwell painting of an inventor puzzling over a perpetual motion machine.

It's easy to see why the idea caught hold. The air was alive with mechanical possibilities. Steam power was driving shipping, sewing machines came onto the market, the first "horseless carriages" and airplanes had become realities. Why not perpetual motion as well? So many patent applications for perpetual motion machines were submitted to the US Patent Office that official policy intervened by declaring that any perpetual motion patent application must include with it a working model. (A policy that still holds true today.)

Despite concurrently developing scientific evidence that perpetual motion could not truly be perpetual, which surfaced as early as 1829 and was solidified by the mid-nineteenth century as the first law of thermodynamics, inventors kept trying their hands at it. Prominent names joined the quest: Sir William Congreve, a pioneer in the field of rocket artillery, developed a perpetual motion machine based on capillary action around 1827. Nikola Tesla claimed to have developed an abstract theory for how perpetual motion might work, though he did not produce a prototype.

Even physicist Richard Feynman proposed a theoretical perpetual motion machine called a "Brownian ratchet" that could convert thermal energy into mechanical work, though he also demonstrated how it failed to work in practice.

More recently, in the late twentieth century, David Jones, a British chemist, inventor, and writer, became well known for producing a series of fake perpetual motion machines, including one held in the Technisches Museum, Vienna. Intriguingly, however, Jones produced one such machine that has continued to baffle scientists even after Jones's death in 2017. In 1981, he displayed a constantly rotating bicycle wheel, with no power source, sealed in a plexiglass container. Jones claimed it was a "scientific joke"; however, no one has yet discovered how the device works. So the joke's still on us, more than 40 years later.

When Jones, who had a flair for the dramatic, knew he was dying from cancer, he wrote down the secret of the wheel's operation in a letter to his friend, Sir Martyn Poliakoff, a chemist at the University of Nottingham. The press tracked Poliakoff down shortly after Jones's death to ask whether he would reveal to the world how this wheel appears to be violating the laws of physics. Poliakoff opened the sealed envelope on camera, with BBC reporters watching. The result was disappointing. Poliakoff said he couldn't quite make sense of it; the details were still baffling to him . . . and then he declared that even if he does eventually work it out, he won't be revealing the secret to anyone.

And so the bicycle wheel continues to spin and spin inside its plexiglass container, inviting marvel and speculation but providing no answers.

David Jones's perpetual motion machine: still baffling people today

World Ice Theory

A **dolf Hitler spoke in grand terms** about building a huge observatory in his hometown of Linz, Austria, in tribute to "the three great cosmological conceptions of history—those of Ptolemy, Copernicus, and Hörbiger."

If you've heard of Ptolemy and Copernicus but are not quite catching the Hörbiger reference, you're not alone. Unless you were a German or Austrian alive in the first half of the twentieth century, you've likely never heard of Hörbiger or his cosmological ideas. And there's a very good reason for that: Hörbiger's ideas were so wildly off base, so deeply incorrect, and so entirely discredited that no one except the Germans under the spell of the Third Reich believed in them. His ideas gained traction because they represented an alternative worldview opposing that of the leading cosmological thinker of the era—a patent clerk named Albert Einstein, who also happened to be Jewish.

Hanns Hörbiger, the would-be astronomer so beloved of Hitler that he intended to dedicate an entire floor of his observatory to him, was the leading proponent of "World Ice Theory," or Welteislehre. An Austrian engineer, inventor, and businessman born in a suburb of Vienna in 1860, Hörbiger helped with the construction of the subway in Budapest and invented a compressor valve still in use today. He also happened to have some eccentric, and rather elaborate, theories on ice.

According to Hörbiger's World Ice Theory, ice was the basic element in all cosmic events and processes, visually evident in our moon (a round ball of ice!) and the Milky Way (a ring of ice blocks!). But for Hörbiger, ice was so much more—it could be used to explain not only astronomical, geological, and meteorological phenomena, but also the entire cultural history of our world. This "kosmotechnische Weltanschauung," or "astronomy of the invisible," as Hörbiger put it, transformed a quack idea on the fringes of astronomy into a full-fledged cultural movement, almost a religion.

Buckle in—here's how Hörbiger's World Ice Theory works: A long time ago, a dead, waterlogged star crashed into our sun, which is the true center of the universe. The massive explosion destroyed the dead star, and the extreme cold of space immediately froze its watery contents. The result: millions of chunks of ice spewing across the galaxy in every direction. A ring of the largest ice blocks would form the Milky Way, while other ice blocks would create solar systems elsewhere in the galaxy. Meanwhile, the planets in our own solar system were significantly affected by the ice explosion. Some of those planets grew larger because they had absorbed so much ice. Others, such as our own, did not subsume as much ice, though they

Hanns Hörbiger not understanding an astronomy textbook, while looking supremely confident

did get pummeled by ice in the form of meteors (also made of ice, according to Hörbiger).

In Hörbiger's icy vision of the cosmos, unhindered by scientific fact or observation, the sun featured prominently as the central gravitational force, slowly, but inexorably, drawing all the planets and free-range ice blocks toward it. This process was aided by the traces of hydrogen gas in interplanetary space, which also caused the planets (and the floating ice) to spiral slowly inward toward the sun.

Meteors were explained away as traveling ice blocks; even hailstorms were the result of small ice blocks colliding with Earth. Sunspots were what we observed when these ice balls hit the sun.

Excessively large ice balls on their way toward the sun could also be "captured" by the gravitational pull of planets they met along the way, becoming moons. Here on Earth, according to World Ice Theory, the moon you observe in the night sky each night is actually the fourth in a long line of moons captured by our planet.

What happened to the previous moons? Hmm, the problem with capturing a moon is that once the moon is captive to the planet's gravitational pull, it's only a matter of time until the moon crashes into the planet itself.

According to Hörbiger, all geological upheavals over the history of Earth could be explained by these damn ice moons ramming into the planet over and over. You can even identify the rock strata of different geological eras based on the impact of the previous moon satellites. Really.

But not.

Hörbiger and his followers were particularly interested in the last moon—the one that orbited Earth before its current moon, which he referred to as the "Tertiary" or "Cenozoic" moon. A British mythology student named Hans Schindler Bellamy, who was a particularly enthusiastic follower of Hörbiger, explained the reasons why: This tertiary moon's capture, and eventual collision with Earth, happened within the time frame that *Homo sapiens* evolved into conscious creatures. Bellamy thought the distant cultural memories of the moon's collision with Earth could be found in surviving legends about human battles with gods and monsters.

By Bellamy's (intuitive) calculations, the tertiary moon was smaller than our current moon. In the cycle of its demise as it drew closer and closer to

• WORLD ICE THEORY | 43

Moons are easy to mistake for dragons, as you can see.

Earth, it spun faster and faster around our planet. The gravitational pull of the moon drew the world's oceans into a so-called "girdle tide," which was apparently a high, narrow belt of ocean close to the equator. This, in turn, pushed the rest of the world into an Ice Age, which forced humankind to seek out "island refuges," that is, high mountain areas near the ocean girdle, particularly in Mexico, Ethiopia, Bolivia, and part of South Asia. Meanwhile, the moon edged closer and closer, eventually circling the Earth six times in a day.

The repetitious, increasingly frequent circling of the tertiary moon, combined with observations of its pitted surface (resembling lizard scales), became remembered as the flight of "dragons" or other monsters in the sky. So all the dragon myths of our cultural ancestors were really misremembered details from when a moon was circling the Earth six times a day.

(Don't judge Hörbiger and his followers too harshly—it's easy to mistake a moon for a dragon. They're both... things in the sky.)

Eventually, the gravitational pull of the Earth was so strong, and the moon so close, that it unleashed a series of catastrophic events. A thick layer of moon ice melted, causing massive rains and hailstones. Next came a torrent of rocks as the moon was ripped apart by the Earth's gravitational pull, leaving the tertiary moon completely destroyed.

Without a moon, the theory held, Earth's oceans were freed from their "girdle tides" and quickly spread in a deluge all over the planet, leading to later legends of a great flood. After the flood, Earth enjoyed a respite from all the destruction. A mild climate and calm oceans were enjoyed the world over. Later legends of an Edenic paradise stemmed from this lull between moons. It was also the time that the city-state of Atlantis arose, becoming an advanced civilization with a global reach.

Sadly, all good things must come to an end, and about 13,500 years ago, the next large ball of ice came floating our way, only to be captured by Earth's gravitational pull and turned into our current moon. Capturing a moon, as our ancestors supposedly observed, is no easy feat. The planetary axis shifted; the poles glaciated. Worse still, the civilization of Atlantis was sunk in all the chaos. Hitler himself believed in just that: The "World Empire of Atlantis... fell victim to the catastrophe of the moons falling to Earth."

Like Atlantis, we Earthlings must also prepare for the logical inevitability and apocalyptic warning that Hörbiger left us with—in short, that it's only a matter of time until our current (fourth!) moon crashes into Earth one day.

Science Shmience

How did Hanns Hörbiger arrive at his elaborate theory? Not through scientific evidence, as just about any kid with a basic understanding of our solar system will tell you. Instead, Hörbiger leaned on that most reliable harbinger of scientific breakthroughs: dreams.

In what was to Hörbiger a particularly revelatory one in 1894, he was floating around in space when he observed Earth as a pendulum suspended from a thread. Gradually, it swung in longer and longer arcs, first to the orbit of Jupiter, then to Saturn, then beyond. But when it swung to three times the distance of Neptune, the thread broke. At this point, Hörbiger awoke, with revelation rushing through his veins. He later concluded, "I knew that Newton had been wrong and that the sun's gravitational pull ceases to exist at three times the distance of Neptune." He also was convinced that most geological and astronomical sciences could be explained through the interplay of "antagonistic Ur-substances of ice and fire."

Even by the loose standards of crackpot dreams, the conclusions Hörbiger reached about the Earth-as-a-pendulum are a stretch.

But for Hörbiger the dream solidified what he had already suspected from his childhood observations of the moon with a small telescope. To him it looked curiously, suspiciously, like ice. After all, it was round and white. When he turned his telescope over to nearby Venus, well, it also looked like ice. And the stars—they, too, looked a lot like ice.

Unencumbered by even the slightest amount of astronomical training, Hörbiger set out to find, or fabricate, the evidence to corroborate his intuitive understanding of the cosmos. Enlisting the support of an amateur astronomer, Philipp Fauth, Hörbiger eventually published his Welteislehre masterpiece, *Glazial Kosmogonie* (*Glacial Cosmogony*) in 1912. The 772-page doorstopper (printed in two columns, with 212 illustrations) is one of the great classics of pseudoscience. As Martin Gardener described it in *Fads and Fallacies in the Name of Science*, it is "filled with photographs and elaborate diagrams, heavy with the thoroughness of German scholarship from beginning to end, [and] totally without value. It is almost as though the Germans, so superior in most fields of scientific learning, refused to be surpassed even in the field of pseudoscience."

The publication of *Glazial Kosmogonie* served as the high-water mark for what German scholar Christina Wessely refers to as the "first period" in the history of World Ice Theory. During this time, Hörbiger and his still-small following attempted to reconcile his intuitive vision of the theory with artificial experiments that might prove its truth. Attempts to spread the word about World Ice Theory in this period largely failed, however, as scientists dismantled it with extraordinary ease.

But Hörbiger was nothing if not determined. Unimpressed with the scientific dismissal of his World Ice Theory, he decided to try a new approach in the aftermath of World War I—he took his idea directly to the masses. And so began the "second period" in the history of World Ice Theory, which stretched from the end of World War I to the rise of the German Nazi party in the 1930s. Hörbiger theorized that if the masses accepted his idea, they would be able to force World Ice Theory into academic discussions.

Sadly, it worked.

"Out with astronomical orthodoxy, give us Hörbiger!" was the cry of Welteislehre proponents as they interrupted educational meetings. Supporters maintained an Information Bureau in Vienna, a monthly magazine called *The Key to World Events*, a handful of hard science books "proving" the theory, a host of popular novels demonstrating it, and a gluttony of ephemera and throwaway pamphlets. World Ice organizers launched "cosmotechnical" societies in German and Austrian cities, hosting public lectures that attracted up to 1,200 people. There were even movies made in support of Welteislehre.

But why was Welteislehre so popular? It was a case of right theory, right place, right time. The burgeoning rise of nationalistic socialism in Germany coupled with its anti-intellectual, mystical worldview made this homegrown cosmological theory extra appealing. It contained a "fascinating amalgam of scientific terminology and methodology with popular images and cliches," according to Wessely. Standing in opposition to mainstream science, Welteislehre also deliciously combined fantasy and reality in a way that was rewarded with extraordinary enthusiasm.

With a long white beard, a fierce gaze, and an unshakable faith in his own convictions, Hörbiger was the perfect leader for the movement. A cult naturally formed around him. As the supreme leader and figurehead of the movement, Hörbiger's word was gospel, allowing him to quickly change the logic of the theory in response to criticisms. Visual evidence that debunked his theory was uniformly dismissed as "fake." (Sound familiar?) He also viciously attacked his detractors. (Also sound familiar?) When German American rocket authority Willy Ley wrote to Hörbiger and pointed out that surface temperatures on the moon had been measured in excess of 100°C (212°F) in the daytime (which would clearly melt ice), Hörbiger tersely and rather ominously wrote back, "Either you believe me and learn, or you will be treated as the enemy."

Ignatius Donnelly and Ragnarok

Ignatius Loyola Donnelly (1831–1901) was a fiery Irish-American populist and writer who believed that a comet hit Earth 12,000 years ago, destroying the advanced civilization of Atlantis.

Donnelly was obsessed with the idea of Atlantis, penning a famous and successful book on the supposed civilization published in 1882. The next year he published *Ragnarok: The Age of Fire and Gravel*. In it, Donnelly laid bare his claims: An enormous comet slammed into the Earth 12,000 years ago, causing a global environmental catastrophe. Fires raged, floods were rampant, poisonous gases were released, and winters became excessively long and difficult. The advanced civilization of Atlantis was also destroyed, and survivors were forced to seek shelter in caves, where all knowledge of their arts and advanced technology was lost over the ensuing centuries.

For proof, Donnelly turned to comparative mythology, citing references to a cataclysmic event from cultures as disparate as Pictish and Hindu, Zoroastarian and ancient Greek. Scientific proof, such as it was, came from the deposits of unsorted, unstratified clay and gravel that scientists describe as "glacial till," left behind by retreating glaciers in the wake of the last ice age. Not so, for Donnelly. According to him, what you may know as "glacial till" is really the dust left from the great comet's tail. In addition to being a fringe scientist and promoter of Atlantean theories, Donnelly also served in elected office. No matter his dabblings in pseudoscience, he was elected both as a US congressman and a lieutenant governor for the state of Minnesota.

Either my aquarium, or Atlantis in all its antediluvian glory

Promising the Moon

Although he was trained as an engineer, Hörbiger placed no value on the scientific method or mathematical calculations. Indeed, he was enormously proud of never calculating anything. "Instead of trusting me, you trust equations!" Hörbiger said to detractors foolish enough to question Welteislehre. "Calculation can only lead you astray."

Indeed. Mathematical calculations would need to be discarded entirely if we were to believe that the sun's gravitational attraction stops at three times the distance to Neptune.

German scientists initially picked the Welteislehre theory apart, which was easy to do because it's so ridiculous. The Austrian astronomer Edmund Weiss pointed out that Hörbiger's intuitive methods of scientific explanation could just as easily be used to argue that the cosmos was made of "olive oil." (Weltoliveollehre?) But the theory was quickly and passionately embraced by the people of Germany and became a movement unto itself in Austria. Millions of actual people bought into Welteislehre, becoming passionate supporters of an astronomical idea that was increasingly dressed in the clothes of a political and cultural movement.

Hörbiger not understanding something . . . again

Inevitably, craven political parties see opportunities in popular movements and embrace the ideas as their own, as was the case with World Ice Theory. When the anti-intellectual National Socialists came to power in Germany in 1933, they adopted World Ice Theory as their own, launching the "third period" in its history (though Hörbiger wouldn't live to see it; he died in Vienna in 1931). But his theory would sadly become the perfect homegrown Germanic alternative to the burgeoning "Jewish theory" of relativity.

The largest supporter of Welteislehre within the Nazi party was Heinrich Himmler, leader of the SS, who was convinced of the theory's truth. "The Aryans did not evolve from apes like the rest of humanity," wrote Himmler, "but are gods come directly from Heaven to Earth." In his view, the Aryans emerged from "living kernels" conserved in "the eternal ice of the cosmos." To him, German Aryans were cosmic gods birthed from space ice. (You really can't make this stuff up!)

Himmler would lead the effort to sponsor Welteislehre as a state science shortly after Hitler achieved a dictatorship over Germany in 1934. By the following year, Himmler had brought together the leading theorists into a quasi-Welteislehre department at the Ahnenerbe, the ancestral research institute set up by the Nazis to popularize "relevant research findings among the German people"—that is, an official conduit to disperse its radical theories and pseudoscience.

The Führer himself embraced the idea, which he believed could one day ultimately replace Christianity. It's no wonder that he wanted an entire floor of his planetarium to be exclusively dedicated to Welteislehre.

As World War II raged, everyone in Germany and Austria had to believe in Welteislehre, along with every other Nazi doctrine, whether they wanted to or not. "Our Nordic ancestors grew strong in ice and snow: belief in the World Ice is consequently the natural heritage of Nordic Man." So claimed Welteislehre literature produced under the Nazis, later quoted by the German American science writer Willy Ley.

Ley quoted another revealing tidbit from Welteislehre publications: "The Führer, by his very life, has proved how much a so-called 'amateur' can be superior to self-styled professionals; it needed another 'amateur' to give us complete understanding of the universe."

And so the doctrine of ice became shamefully intertwined with the doctrine of hate. Thankfully for the world at large, both doctrines came to an end when Soviet troops took Berlin in 1945.

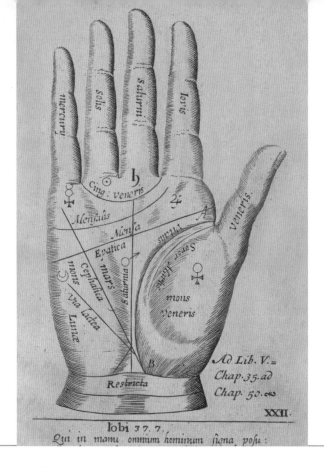

Body Divination

"**Body Divination**" sounds like the newest miraculous luxury skincare product... but it's not. It's something quite old, actually. Long before neon lights drew us to palm reading rooms and friends offered us tarot readings over cocktails, for millennia we attempted to scry the future in every possible way. And not by looking at the stars, but by examining something much closer: our bodies.

People love to find patterns and messages in everything: thrown sheep knuckle bones, the constellations, glass orbs, tea leaves. The list is quite enormous—and overwhelming. There's alectryomancy, or using roosters for prediction; scatomancy, which uses excrement; and, of course, there's kephalonomancy, which uses a donkey's head. And don't forget phrenology (see page 64). Or pillimancy, a way of ascertaining your future through the tangles of hair that fall out of your head.

And while there's some medical evidence for a few methods of divination (see below, not above), most approaches to body divination are clearly quackish. In no particular order of importance, let's see how the marvelous-yet-ordinary human body can supposedly help us journey through the looking glass and see the future.

Bottom's Up!

Rather than beginning up top, it's better to start down below—that is, with the buttocks. The tush. The derrière. This type of divination is called "rumpology," a term reportedly coined by actor Sylvester Stallone's mother, Jackie Stallone, who was well-known for her psychic work, including writing astrology books, running a hotline for psychic readings, and reportedly being able to tell fortunes by using dogs. But rumpology was probably the most infamous of her divination practices. She claimed the method was used in ancient Babylon, India, Greece, and Rome. It's a common practice used by con artists—referring to antiquity as a means of foisting historical legitimacy onto your work—though it isn't always a selling point. After all, there are plenty of ancient-world practices that are considered outright awful by today's standards (think: medicinal cannibalism or using burning hot irons to cure lovesickness).

To really get to the bottom of how rumpology works, you need to understand how a reading is done: You can either make a print of your glutes (butt cheeks) by painting them with ink or henna, then pressing them onto a

Jackie Stallone. She can't wait to get a handle on your . . . future.

piece of paper to make an impression, sort of like an enormous bilobed fingerprint ... or ... simply drop your pants for a direct reading. Nowadays, a Skype or Zoom rump reading would allow for a visually accurate yet socially distanced examination. But some readers might require a more, shall we say, tactile experience, where the rumpologist actually feels the bumps, dimples, and fleshiness of your booty.

Supposedly, the left cheek represents the past, so the right cheek is obviously a predictor of the future. (This is sometimes reversed by the rump reader, if appropriate. Why reverse it, you ask? The better question is why introduce logic into the concept of rumpology at all?) The "crack," or gluteal cleft, has meaning, too. Long, short, deep, shallow—the characteristics are all part of the reading. Jackie Stallone reportedly opined, "It represents the division between the ying/yang, good/bad, light/darkness, between your past ... and your future ... Many banker's clefts are very short; while lawyers are very long."

Oh, and your posterior prognosis will cost you. Jackie Stallone charged $300—per cheek.

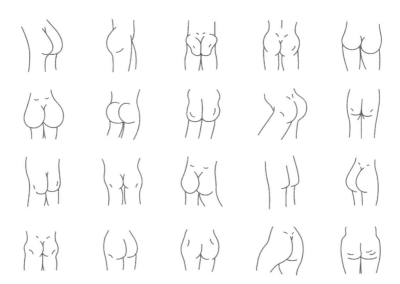

A variety of futures for your perusal

But the mother of Sly wasn't the only fanny fanatic. A blind German clairvoyant shaman named Ulf Buck does rump readings in his tiny village just north of Hamburg. During a spiritual retreat around 1999, he had visions that inspired him. "The most vivid images were of naked buttocks," he said. "I could see clear horizontal lines running along each hemisphere . . . I realized I was looking into the future."

Some lines on the buttocks foretell success or failure in real estate. Or in relationships with family. Shapes might also point to certain personality traits—for example, those with apple-shaped, muscular butts may possess charisma, confidence, and creativity. Another rumpologist, Sam Amos of England, thinks that "a flat bottom suggests the person is rather vain and is negative and sad." No word yet on how a surgical Brazilian butt lift might alter your fate.

Look Me in the Iris!

From the bottom to the top, some might prefer to see the future in their own eyes—the so-called windows to the soul. Called iridology, this bodily prediction method purports that a person's health can be seen in their irises, or the colored discs surrounding the pupil. Not only do iridologists claim to diagnose current problems, but they also supposedly have the ability to foretell future health challenges as well.

The practice of iridology is based on a complex chart that divides the iris up like a clockface or a pie, with each thin slice corresponding to a different part of the body, such as the pancreas, penis, thyroid, or throat. To get a really close look at the iris, it often needs to be examined using a slit lamp, commonly found in the offices of optometrists and ophthalmologists.

Like myriad other quackish approaches, iridology actually has a long history that can be traced. Since about 3000 BCE, healers and medical practitioners looked at our eyes for clues of illness. But it wasn't until the nineteenth century that a term for diagnosing disease by examining the irises, "iridology," emerged. Hungarian physician Ignaz von Peczely, regarded as the "father of iridology," once noticed that a man with a broken leg and an owl with a broken leg had something in common—they both had the same streaks in their eyes. He concluded that their eyes spoke to their

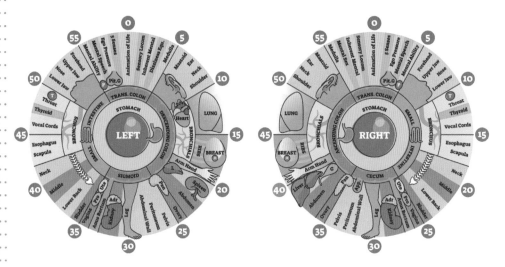

Gaze into my eyes and you shall see . . . my rectum?

bodily diagnoses, possibly before the disease even showed up. He wrote a book on the subject that others in the field soon picked up, while practitioners in Europe continued working on the theory.

The problem is irises don't change much at all during our lifetimes. In fact, it's precisely why eye scans are sometimes used as a biometric identity feature. There are certain health conditions that show up in the eye, of course, which can lead to a diagnosis. Wilson's disease, a disorder where copper accumulates in the organs of the body, may produce a telltale copper-colored ring overlying the iris. Brownish nodules appear in those with a hereditary disorder called neurofibromatosis. And a type of skin cancer, melanoma, can present with a new, dark freckle on the iris. But most of the time, our irises don't change.

Several reviews and studies showed that iridologists can't predict diseases or failing organs accurately using the iris. In one 1979 study, three leading iridologists were unable to diagnose which patients had normal or severe kidney disease based on their irises. A *British Medical Journal* article in 1988 gave iridologists a chance to diagnose those who had inflamed gallbladders, but their predictions weren't much more accurate than basic chance.

It's Only Skin-Deep

But if it turns out our eyes aren't the way to see into the future (or the present, for that matter), perhaps our skin is. After all, who hasn't heard tales of people having a revealing mark on their skin, a sign that they were destined to be some kind of savior or devil incarnate? In David Mitchell's *Cloud Atlas*, reincarnated people always carry a comet-shaped mark on their skin. Some cultures believe that a child's birthmark occurs due to an unfulfilled craving of the mother during pregnancy.

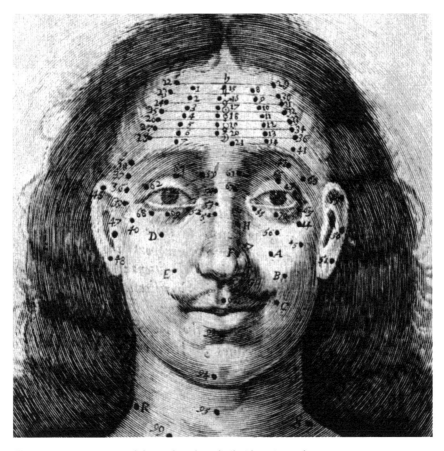

First rumps, now bumps. A future is written in that beauty mark.

In addition to odd-shaped birthmarks, moles were also used to glean information about a person (called moleosophy, not to be confused with the study of underground rodents). For this history, we go back to Greek antiquity, where the prophet Melampus wrote "Peri Elaion tou somatos," which roughly translates to "About body olives" (olives because they might have the same texture, color, and shape as moles). In an English translation by Tim Spalding, Melampus writes, "If a man has a mark on his chin (or wattles), he will be rich in gold and silver. The same also for a woman, when she has one on her spleen." Also, this dire prediction: "If a man has it on the back of his throat, he will be beheaded." And this very particular one: "If it is on the nose of a man, and his skin is ruddy, he will be insatiate in lovemaking, also when she has a birthmark also upon the privates."

Fingernails were further observed to be useful in the act of divination. Called onychomancy, this practice usually required the nails to be coated in oil (and sometimes soot), then observed under strong light, such as sunlight. Apparently, a young boy's nails were most frequently used to portend the future (whether of the boy or someone else is unclear). However, polymath Edward Heron-Allen wrote extensively of the hands in his extravagantly named 1885 book, *A Manual of Cheirosophy: Being a Complete Practical Handbook of the Twin Sciences of Cheirognomy and Cheiromancy, by Means Whereof the Past, the Present, and the Future May Be Read in the Formation of the Hands Preceded by an Introductory Argument upon the Science of Cheirosophy and Its Claims to Rank as a Physical Science.*

Heron-Allen wrote that the medieval methods of oil-polished fingernails for auguries was "too absurd." But he did think the nails indicated character. Apparently, short nails denoted "sharpness, quickness of intellect, and perspicacity." Broad nails meant you're "pugnacious, critical in disposition, fond of domination and control." Long and curved nails indicate "ferocity and cruelty." White marks on the thumbnail denote "affection, which is generally reciprocated." A white mark on the forefinger nail means a gain; a black mark "marks a loss." Aside from Heron-Allen, other fingernail readers associated the different parts of the nail bed with the different energy points in the body, or chakras, a term that originates from Hinduism and Buddhism.

Talk to the Hand

Far more famous than fingernail fortune telling is palmistry. Also called chirology or chiromancy, palm reading involves studying the lines and other features of the palm to interpret the futures and tendencies of its owner. Podomancy, or readings of the foot soles, was far less popular throughout history. (One can imagine that the smell, and the general state of the average person's toenails, might have made the latter less appealing to the readers.)

Chiromancy is an ancient technique found in many cultures. Believed to have gained prominence in India around 2500 BCE, it was mentioned in ancient Indian Vedic texts, and likely spread beyond India to China, Sumeria, Babylonia, and other Eurasian cultures. Palm readers in the Romani culture, with its roots in northern India, likely draw many of their soothsaying traditions from the rich history of chiromancy. Palm reading arose independently in ancient Greece around the time of Aristotle in the fourth century BCE.

Biblical verses also mention palms as a source of fortunes, as in Proverbs 3:16, where the "length of days is in her right hand; and in her left hand riches and honor." In his *Historia Animalium*, Aristotle mentions that a person's longevity is revealed in the palm. Those with long lives have two long lines extending across the palm, compared to short-lived people, whose two lines don't extend across the entire palm. (For the record, he also believed that the brain was a sort of radiator that cooled off a temperamental "hot" heart.) Thus, the palm reveals a smaller version of the whole body. It's a concept that is seen throughout the practice of body divination, or of finding larger truths within the smaller aspects of the human body.

A Tibetan fortune teller reads the palm of a Nepali man.

Fingernail Cutting Superstitions

Everyone has nails that need clipping. Unlike birds and wild horses, our hoof and claw equivalents need to be actively snipped off, unless we're attempting a Guinness World Record. But be wary of when you choose to wield those clippers. It could decide your fate.

An old English rhyme from *The Home Book of Verse* in 1912 shares this superstition:

Cut your nails on Monday, cut them for news;
Cut them on Tuesday, a pair of new shoes;
Cut them on Wednesday, cut them for health;
Cut them on Thursday, cut them for wealth;
Cut them on Friday, cut them for woe;
Cut them on Saturday, a journey you'll go;
Cut them on Sunday, you'll cut them for evil,
    For all the next week you'll be ruled by the devil.

A Korean superstition holds that you should not cut nails at night, for fear that rats will eat the nails. With bellies full of your discarded body parts, the rats would then be able to transform into you and steal your soul. Horrifying!

An obscure medieval manuscript, attributed at the end of the document to Aristotle but clearly not written by him, makes all sorts of revelations about our hands. A large, fleshy one means its owner drinks too much alcohol. Broad fingers with sharper tips mean a person is lusty and prone to lying. Most hand readings for women tend to reflect upon their fecundity. Broad fingers mean she'll have a voluminous womb and will be on the hunt for a man. A small "table line" (now often called the "heart line") but large fingers mean she's headed for the sex industry.

Medieval mathematician and scholar Michael Scot believed that to find out the gender of an unborn child, you simply ask the pregnant mother to hold out a hand. If she extends her left hand, the child will be a girl. If she

In Japan, nighttime nail cutting draws death ever closer. There is a saying that goes, "Yoru ni tsume wo kiru to oya no shinime ni aenai," or "If you cut your nails at night, you won't be with your parents when they pass away." The belief may have arisen from the word "yotsume" (夜爪), which means "night nails" and is pronounced in a similar way to another word (世詰め), which means "the end of the world" or "short-lived."

In Hinduism, some believe that nails should not be cut on Saturdays, which is perhaps related to the deity Shani, the personification of the planet Saturn, who wields great power over human life and longevity. Saturday is a day of worship for Shani, who would be angered if you trimmed a piece of yourself off on that day. And apparently, cutting nails at night would anger the goddess Lakshmi. Wednesday—during the day, of course—is a much more auspicious time for clipping. Overlapping with secular Indian culture, many people in India follow a habit of not cutting nails at night, for fear that doing so would bring sickness and bad luck.

Why all the nail-paring fear? Some believe it stems from preindustrial times, before artificial bulbs illuminated our homes at night and we lacked those safe, nifty nail clippers that are ubiquitous today. It's easy to imagine what a bad idea it would be to wield a knife in dim light. Doing so was asking for an awful accident, not to mention a deadly case of tetanus.

But not angering all-powerful deities is a pretty strong motivator, too. Especially when it's as avoidable as doing the deed on a different day and in broad daylight.

offers her right hand, it will be a boy. So there's a 50 percent chance of getting it right, though let's be honest: You don't need to see which hand a pregnant woman extends to get those odds.

Toward the end of the nineteenth century, chiromancy became yet more wildly popular, as did many types of occultism. Chirological societies popped up, and preeminent Western chirologists published their knowledge, many of which are still used as guide texts today. One was an Irish astrologer, William John Warner, also known by his stage name, Cheiro. According to his memoirs, Cheiro developed his craft while traveling to India on a trading vessel, where he encountered some Brahmins who specialized in Samudrika, or the practice of reading the lines of the body. Soon he had a roster of

William John Warner (aka Cheiro), one of the first celebrity palm readers

famous clientele, including Mark Twain, Oscar Wilde, Thomas Edison, and Sarah Bernhardt. For Bernhardt, he said he saw in her palm "an active determined will" and an "artistic emotional career," and the reading left the pioneering French actor in tears. His reading for Oscar Wilde foretold that "the right hand is the hand of a king who will send himself into exile." When asked by Wilde when this would happen, Cheiro stated, "A few years from now . . . between your forty-first and forty-second year," which was when Wilde was later incarcerated on the conviction of gross indecency for having sexual relationships with men.

An esteemed historical chirologist, William Benham published a thorough handbook at the turn of the twentieth century claiming a divine reasoning for palmistry. Benham said, "He made the human hand a reflector of the brain, and He wrote in it a language plain, simple, and easy to understand, which He intended that we should use for our benefit, for He put in human brains the key by which this language could be read."

It sure seems like a good idea, except . . . can palms truly predict who you are, or what your future is? Let's consider the evidence.

All Fingers and Thumbs

A key difference first: Palmistry is not dermatoglyphics. The latter is a scientific study of things like fingerprints, creases of the hands, mounts (the fleshy areas between lines), and hand and finger length as they pertain to certain health conditions. Dermatoglyphics is actually used in the medical and forensic world. Think law enforcement identifying people by their fingerprints, or doctors observing unusually deep creases of the hand that may point to a genetic finding, like Down syndrome. About half of people with Down syndrome have a single, transverse palmar crease. People with high blood pressure tend to have whorl patterns on their fingertips.

Your fingerprint, actually

Is your ring finger longer than your index finger? If you're a man, it might mean that you have a more attractive face, a longer penis, and better athletic skills. Preschoolers with a longer index than ring finger have a better ability to delay their gratification and wait patiently for a marshmallow or pretzel treat, as shown in the famous Stanford marshmallow experiment. Why? It's not clear, but there are some studies that show that in utero exposure to androgen hormones can affect finger length and create preschoolers who aren't so great at waiting for their treat.

But can people use these findings to predict poor health? A study in 2020 tested two "academically qualified palmists" to read the handprints of patients with ALS (amyotrophic lateral sclerosis, or Lou Gehrig's disease), as well as control subjects. Their predictions were consistently 40 percent accurate or lower on guessing comorbidity (other medical illnesses) or what organs would cause death. Which isn't really that useful, in a medical scenario.

In another study in 2019, scientists tried to correlate the length of the palm's "life line" with the age of death of 60 cadavers at an anatomical department at the University of New South Wales, in Australia. This followed two studies—one in 1974 that found no correlation, and one in 1990 that said the life line did correlate with age of death. The Australian study didn't find a correlation.

And in any event, most people looking to predict their health destinies are more likely to turn to DNA testing than to the lines in their palms. What someone seeking a palm reading wants to know about isn't the health of their spleen. Rather, they're looking for an interpretation of what the lines and fleshy bits of their hands say about whether they'll fall in love this year, or perhaps lose a huge amount of money.

Cold Comfort

We should probably tackle the elephant in the divination room, because there's an enormous factor that influences a palm or butt reading. Hint: It's not your palm or your butt. It's not even money.

Enter the "cold read."

Cold reading involves a person—a psychic or a palm reader, for example—who savvily absorbs everything about you before you even extend your hand, which will generally first be extended with a credit card for payment. They look at things like how you're dressed, how expensive or worn-out your clothes are, what jewelry you're wearing and if it's subtle, genuine, nonexistent, garish, or includes a set of 50 titanium body piercings. They'll notice if you're shy or chatty and try to ascertain your level of education. By way of these methods, a fortune teller or clairvoyant gets to know you without having looked you up on Google beforehand (but chances are, they will if they can).

As a matter of practice (and preamble), they might talk about the long, illustrious history of palm reading and how accurate it is—this is to prime your belief in what they have to say. They might also educate you ahead of time about the lines on your hand and how complicated they are to read. They'll also likely talk about how many years they've been doing palm reading and the many times when they have been correct. But they may also be quite humble, keeping the claims subtle or not promising too much, to keep your expectations reasonable and not give off quacky salesperson vibes. This could include talking in gray zones and asking "fishing" questions that lead them to more information. "Was there a disappointment in your life recently?" "I see a person—someone you know a little but want to know better." "It has something to do with your job. Or perhaps a hobby." They might look at your hand and say, "Wow, you have a lot going on!"

Because who doesn't?

Another tactic is to have a list of phrases that seem very complex and rich—platitudes they've probably seen in manuals or learned from other readers. But above all, they'll watch you carefully as they read, adjusting to your excitement, discomfort, or embarrassment. They'll be great listeners no matter what. And they'll probably flatter you. "This hand says so much!" "You've done so many amazing things in your life!" These will surely include some truisms that make sense for everyone. "Wow, you've seen a lot of things in your life." "You're keeping a secret." "To thine own self be true."

Yes, that last one is Shakespeare, but it could easily be read on a palm, as far as advice goes.

· · · · ·

People might lie, but bodies can't. You can tell your doctor that you've never smoked, but fifty years of two packs a day can't be hidden when the emphysema shows up on a scan. You can tell a police officer you didn't drink, but your blood alcohol level might say differently. That being said, the body can't reveal certain things you desperately wish to know, namely your romantic, emotional, work-related, and financial future in detail.

So if you're looking for some advice via a palm reader—or, goodness gracious, from a rumpologist—be ready to pay for an experience wrapped up in a package of pseudoscience. It might be fun, but it won't be accurate unless by chance, and you might be better off saving your money and consulting a wise friend instead.

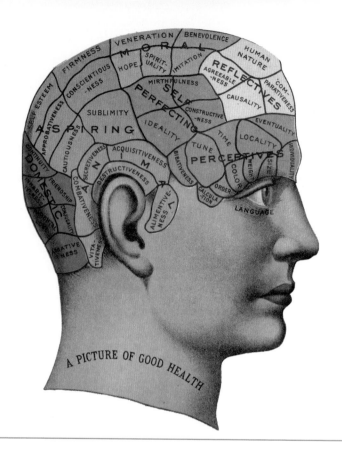

Phrenology

You might think that someone with, say, large biceps would be pretty good at lifting things. Fair enough. Or that someone with large, fine hands might excel at playing the piano. We'll buy that.

How about eyesight? The more bulging and large the eyes, the better they would be at visual acuity, right? Or people with larger noses—they can smell scents better, no?

Here's where anatomy and the common sense of correlations disintegrates. If you already figured that out, you're an expert at phrenology. That is to say, you know it's a bunch of quack science.

"Phrenology," from the Greek words for "mind" (phren) and "knowledge" (logos), is the study of human behavior and aptitude based on the measurements of the skull. The basic tenets went like this: If certain parts of the brain were particularly well developed, then it affected the overlying skull. Which essentially meant that you could figure out a person's sexual appetite, eating habits, genius or stupidity, penchant for lying and criminal activity—all based on the bumps of their head.

Before you start massaging your scalp looking for the I'm-gonna-be-rich-and-famous bump or the Holy-crap-I'm-screwed-for-life bump, let's take a closer look.

Low-Brow, High-Brow

This story begins with Franz Joseph Gall, the father of phrenology and an eighteenth-century Viennese physician. Gall was a skilled anatomist whose interest in the human body began as a child. One of his early observations was that some of the best students around him had bulging eyes. Surely, he thought, they must have enlarged brain matter shoving those eyes forward. Hence, verbal memory must be located in the frontal lobes, creating Gollum-like eyeballs.

What followed was an idea that the brain wasn't one uniform organ, but a patchwork quilt of specialized areas. Unlike in previous millennia, when it was thought that mental activity was situated in the heart, Gall inferred that the mind was in the brain. When one area was stronger, it was bigger and thus caused the skull to rise in that area with a palpable bump. Twenty-seven bumps, to be exact. Nineteen of

Franz Joseph Gall examines a very bumpy head. (While staring at you uncomfortably.)

I Like Big Brains and I Cannot Lie!

Following the work of eighteenth-century phrenologists obsessed with the bumps on our skulls came the nineteenth-century men who desperately wished to prove, once and for all, that they had really frickin' awesome brains.

That's right. People wanted physical proof that their brains matched their awesomeness. In order to do this, of course, they needed brains. In 1824, the proof was in the postmortem examination of Lord Byron, the romantic genius who possessed a massive five-pound brain, two pounds more than the average. Beethoven died in 1827, and the ridges and folds on the surface of his brain (the gyri) were reported at his autopsy to appear "twice as numerous and the fissures twice as deep as ordinary brains."

Around 1855, physiologist Rudolph Wagner found that common laborers sometimes also had big brains. Plenty of ordinary people were revealed to have complex surface fissures in their postmortems, too. But Wagner's work was pooh-poohed. Meanwhile, Paul Broca would become famous for localizing the speech area of the brain to the left frontal lobe (helpful!), while also theorizing that the shape of white European heads proved they were superior to Black people

PHRENOLOGY CHARACTERS

GOOD WIFE   GOSSIP MONGER

WEAK SPIRITED   CRIMINAL

QUICK TEMPER
                 RESPECTABLE

Random character traits associated with random physical features, according to phrenology

(not helpful or true!). A faction of Broca followers created the Société d'Autopsie Mutuelle (yep, the Society of Mutual Autopsy). Members bequeathed their heads, faces, skulls, brains, "and more if it is necessary" to the cause of anatomic grandiosity.

Multiple groups popped up in Europe and America to do the same. One, the Anthropometric Society of Philadelphia, was lucky enough to get their hands on Walt Whitman's brain.

Whitman was widely known to be fascinated by brains. He'd gone to the Phrenological Museum of Orson Fowler and Horatio Welles in 1849 and had a bump reading. It's not really known if Whitman actually gave consent to turn over his brain for medical research. Perhaps he was in an elite brain society—or not. What we do know is that Edward Spitzka, an anatomist who had studied the brains of people imprisoned and dying, collected data on six scientists and scholars who belonged to the society. But there was a noticeable "oops" in his magnum opus of a final report: It didn't include anything on Whitman's brain.

Why?

It had been dropped (and not from the report).

It's not clear who did the dropping, but preserving brains after death for the sake of autopsy and study is, of course, no easy feat. You can't just pickle them lickety-split by dumping one into a vat of formaldehyde or alcohol. It needed to be perfused, or permeated, with successive soakings of preservative over weeks. Whitman's brain probably hadn't been preserved yet, because it apparently shattered into pieces, not unlike dropping a giant, well-cooked flan on a kitchen floor. Or, according to the records of the Wistar Institute of Anatomy and Biology, where Whitman's noodle would have been kept for posterity had it survived, "the brain was accidentally broken to bits during the pickling process."

Despite its grandiose aims, the Anthropometric Society was a disorganized mess. The records of the society are nowhere to be found. And the brain itself? No one admitted that it likely ended up in the trash to cover up the mistake. Spitzka's loss of Whitman's brain was embarrassing, to put it mildly. He ended up suffering paranoid delusions that ex-incarcerated people he studied were out to get him for removing brains from Sing Sing prison. Following a nervous breakdown he died of a stroke, yet another mind lost among the members of the society.

Well, not completely lost. Spitzka did end up donating his brain for research.

Walt Whitman, pre–brain-pickle fiasco

these qualities were shared with animals, but the specifically human ones included wit/joking, sagacity, mimicry, poetic ability, and religious instinct. Vanity? There's a bump for that. Instinct to kill? There's a bump for that, too. Bibativeness? You've got a love of liquids, so lay off the booze.

Before long, Gall started getting his curious fingers onto as many heads as possible—criminals, psychiatric patients, artists. He also started collecting skulls. One of his most enthusiastic students and followers, J. G. Spurzheim, aided Gall in brain dissections. The kaiser at the time, Holy Roman Emperor Francis II, basically ran them out of Vienna, accusing them of being amoral materialists (those who believed that consciousness and mental faculties reside in brain matter itself).

Others later complained that phrenology was a pseudoscience from the start. Oliver Wendell Holmes Sr. (d. 1894), a professor at Harvard, had this to say about it: "Can you tell how much money there is in a safe, which also has thick double walls, by kneading its knobs with your fingers? So when a man fumbles about my forehead, and talks about the organs of Individuality, Size, etc., I trust him as much as I should if he felt the outside of my strongbox and told me that there was a five-dollar or ten-dollar bill under this or that particular rivet."

An 1886 illustration of parents annoying children by stressing out about their future careers

But Gall was undeterred. He left Vienna to promote himself on tour around Germany in style. And by "style," we mean a carnival act complete with animal and human skulls, plaster casts of yet more skulls, and a wax modeler. Oh, and two monkeys (because every self-promoting scientist needs two monkeys).

Gall wasn't welcomed everywhere. The Austrian government asked him to stop lecturing, fearing people would "lose their heads" and become amoral materialists themselves. Eventually, Gall retired after marrying a young woman (he'd had two wives and several mistresses) and, upon his death, thoughtfully willed his brain to Spurzheim to study. Spurzheim continued Gall's work, adding another eight "faculties" to the list Gall created.

• PHRENOLOGY | 69

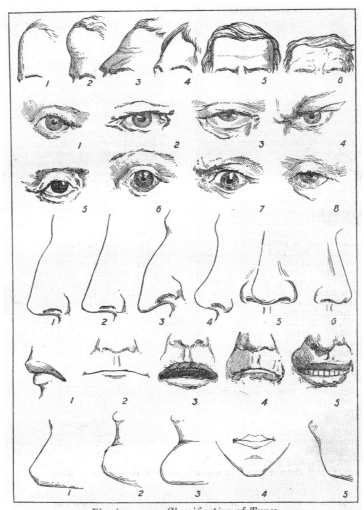

Physiognomy—Classification of Types.

Foreheads.—1, Profound thinker, cold and calculating; 2, quick intelligence and irritability of character; 3, irreflective, despotic, eccentric; 4, imbecile; 5, well balanced intelligence; 6, criminal. **Eyes.**—1, Calm; 2, ardent; 3, apathetic; 4, cunning; 5, phlegmatic; 6, anxious; 7, passionate; 8, indolent. **Noses.**—1, Roman—strong force of character; 2, Grecian—refinement; 3, Jewish—avaricious; 4, snub—weakness of character; 5, negro—secretive; 6, divided—keen perceptive powers. **Mouths** (indicative of)—1, Firmness; 2, coldness; 3, indecision; 4, coldness and cruelty; 5, irritability. **Chins** (indicative of)—1, Selfishness; 2, benevolence; 3, violent love; 4, desire of love; 5, cold nature, lacking affection.

A turn-of-the-twentieth-century illustration of physiognomy, with all its bigoted detail

The number of faculties would prove to be a thorny issue and ultimately one of the downfalls of phrenology. Those darn phrenologists could never agree on how many faculties the brain actually had. (Physiognomy, or the reading of facial features, would also become part of the common vernacular.)

But for a time, people couldn't get enough of phrenology. In a confusing array of contradictions, phrenology was given support from just about everyone, regardless of their point of view on society and anthropology. On the one hand, the upper class thought it solidified the notion that their exquisite noggins and patrician visages with "high-brow" features meant that high-born was better-born. Social hierarchy was therefore considered a good and natural thing, and now backed up by science! Working-class folks sought out phrenology because it held that you had inborn faults and that meant you could improve upon them and rise beyond the station to which you were born. In this sense, phrenology was a precursor to meritocracy.

Phrenology also fanned an already heated argument on racism and slavery. One of Gall's followers, François-Joseph-Victor Broussais, claimed that Indigenous peoples lacked a cerebral organ that created great artists and that Caucasians were the most beautiful. And yet, abolitionists upheld some phrenologists' findings that "the colored man has more natural talent than is generally ascribed to him." Many phrenologists, like Robert Collyer, abhorred slavery, making Broussais's work unpopular with them.

The guise of phrenology made it all too easy to stereotype by gender as well. After all, women were said to have larger skull areas for nurturing and child rearing, so they were naturally built for a life at home. Yet women also attended phrenological lectures and were told by many practitioners that they possessed mental equality with men. In fact, most phrenologists supported women's equality.

Somehow phrenology managed to say what just about everyone wanted to hear (or didn't want to hear), in some fashion. And that message was more or less that the world could be made a better place through phrenology: Criminals could be rehabilitated with the right treatment; societal effort could change people; the insane could be cured.

And where there was a method for improving one's life, there were of course people to offer the perfect solution—for the right price.

A Head for Business

While academics tried to tease out the anatomical and clinical applications of phrenology, others were happy to take the money and run with it.

People in the US especially welcomed phrenology. In a country more or less built on a bootstrapping mentality, phrenology held a tantalizing key to success. And the Fowler brothers knew it.

Lorenzo Niles Fowler and Orson Squire Fowler started their phrenology business in the 1830s in New York City. Orson was a farmer's son who was at first keen on becoming a minister. But he soon discovered that phrenology held everything he wanted—fame, money, and, to a certain degree, respect. He would go on to lecture on phrenology at Amherst College and read heads for two pennies a pop.

Soon, the whole Fowler family got into the business, including Lorenzo; his wife, Lydia Folger Fowler; sister, Charlotte; and her spouse, Samuel Wells. Lydia in particular found a willing female audience. After all, women were more than ready to hear that they had plenty of potential under the bumps of their skulls, and many phrenologists supported women's rights. Lydia received a medical degree from an eclectic school (a type of medical education that consisted mostly of botanical remedies, hygiene, and diet), Central Medical College of Syracuse, and opened a practice in New York. At a typical exam by one of the Fowler family, a customer's head was felt for an hour or so, the examined bumps each given a corresponding number and recorded on paper (the higher the number the better the human potential). The Fowlers established a corporation whose motto was, perhaps ironically, "Know thyself."

Not content to just hawk their readings and phrenological wares, the Fowlers leaned in even harder to the phrenology frenzy. They published their own journal and countless pamphlets and books, selling tens of thousands of copies each year. They were more than happy to tell people how to live, what to eat, how to maintain their marriages and raise kids. During the Victorian era, the phrenologist became your best friend—they'd tell you who to hire in your home, or whether someone was good marriage material. Phrenology was discussed well beyond the circles of upper-echelon

Anyone can be a phrenologist, for a price.

academics. It found a mass audience because it was both entertainment *and* practical advice for everyone.

One advertisement, circa 1838, read almost like a warning for worried parents. "Phrenological Predictions. — Persons meditating any important change in their pursuits, parents, before deciding on a business or profession for their children, should consult this science, as their fortunes depend on the choice harmonising with its predictions. Terms, five shillings and upwards."

Soon, it was all too easy for people to turn a corner and see someone to "get their head examined." Poorly developed characteristics were thought to be malleable, and one could improve one's inborn tendencies for criminality, for example, with effort. Phrases like "well-rounded" stem from positive phrenology traits, such as described in an 1896 book where "The high and well-rounded forehead, indicates poise of intellect; a mind stored with experience." The popular belief still exists that psychiatrists and psychologists are called "shrinks" due to phrenology and its promise to inhibit certain personality traits found in prominent bumps (though phrenology has zero influence on the practice of psychiatry).

The phrenology industry made a lot of practitioners quite a pretty penny selling books, busts, charts, and skulls, in addition to readings. And those poorly trained in it were hard to distinguish from those with what was considered "proper training." Phrenology quackery existed, if you can believe it. The US was apparently filled with "humbug phrenologists" and "unprincipled quacks," according to a follower of Spurzheim, Robert Collyer. Still, Collyer himself also attracted a good deal of side-eye from academics for practicing phrenomagnetism, whereby he would put patients in a mesmeric trance and touch head bumps, allegedly to manipulate them into bigger or lesser effects. "Legitimate" phrenologists only did a reading of your bumps, by contrast. Collyer might touch the bump correlating to self-esteem, giving the patient an arrogant expression. Phrenomagnetism would become an embarrassment after mesmerism was attacked by medical journals in the 1830s and '40s.

Science, it seemed, was the only thing that could finally stop a field that promised a panacea for virtually every problem. By the 1840s, the practice of phrenology was soundly debunked, though it remained popular for decades and its verbiage resonated to the point where it's still interwoven with our contemporary language to this day.

Getting Your Head Examined

By the twentieth century, phrenology had long since lost its luster. But the power of a new invention can suddenly make bad science look good again. And phrenology found a resurrectionist in Henry C. Lavery of Superior, Wisconsin.

A deep believer in the lost "science," Lavery created a device in 1905 that would take the guesswork out of phrenology. It was called the Psycograph—and it would both measure a person's head shape and spit out a written assessment they could take home.

The Psycograph was a metal helmet fitted with probes that touched parts of your head. The sensation it delivered ranged from ticklish to prickly. The probes then turned on a motor that received low-voltage signals from them (they measured size, not electrical brain waves) and printed out an assessment ranging from "deficient" to "superior" in 32 faculties.

So you or your large-headed friend, or a grapefruit, or the neighbor's gigantic garden zucchini could each wear the helmet and get a printout akin to a really detailed fortune that would blow a Chinese restaurant fortune cookie out of the water. A typical subheading might say, "Suavity—Average—You can be pleasant, polite, and tactful with others, but with many people you achieve more by exercising more diplomacy and courtesy." That's right. It was also apparently an asshole assessment.

Customers were further given a vocational chart to help them find a career, such as "zeppelin attendant." The Psycograph reading would then tell them where their strengths were. Put the chart and the reading together, and you could find your best-suited life. We (your dutiful authors) looked up our professions and found that they do not require "suavity" (see asshole comment above), "secretiveness" (propensity to be obnoxiously frank), and "sexaminty" (love life is severely unimportant). So apparently the Psycograph can cause self-loathing, too.

The machine showed up in cinema lobbies and department stores with signs urging folks to "Get Your Head Examined." But despite earning an astounding $200,000 at the World's Fair in Chicago in 1934, the Psycograph's popularity didn't last. (Turns out, it's hard to make a living on a science that most people have already known is a joke for more than 50 years.)

Phrenology is now a thing of the past after scientists thoroughly examined its bumps and charts, tossing them aside for things like solid neuroscience, psychology, and psychiatry. Its lasting influence might be its kitschy value; luckily for everyone, ceramic phrenology busts can still be found in flea markets and home décor websites for a very affordable, low-brow price.

The Psycograph in action, circa 1931

Aliens! Ghosts! Bigfoot! Atlantis!

UFOlogy

On November 11, 1987, a contractor in Gulf Breeze, Florida, had a terrifying experience. He saw an unidentified flying object in the sky that shot a blue beam toward him. The flying saucer was taller than people usually imagine when they think of UFOs, and this one managed to immobilize him for a short time. He was then somehow miraculously able to fetch his Polaroid camera and take a few photos afterward. The visitation victim, Ed Walters, described the experience and sighting as being "straight out of a Spielberg movie."

And then it happened again in December. Walters recalled, "Something hit me. All over my body. I tried to lift my arms to point the camera. I couldn't move them. They were blue. I was blue. Everything was blue. I was in a blue light beam."

To prove he wasn't making this up, he submitted his photos to the *Gulf Breeze Sentinel* under the guise of a "Mr. X." The photos didn't look like flying saucers from early, grainy sci-fi movies. They looked more like puffed-up, blobbish cakes that were brighter at the base than the top. Later, his wife witnessed the UFO, too. They claimed they could hear the extraterrestrials speaking in their minds and it was messing up their lives.

The Walterses weren't alone. Dozens of other people in the area reported sightings, too. But what made the other sightings different from the Walterses' was this: the "four plastic foam plates and some drafting paper" found in the Walterses' home in Gulf Breeze. It turned out that Ed Walters's eyewitness photos were likely double exposure prints of the plates. The fuzzy images were later debunked as fakes by a member of NASA's Jet Propulsion Laboratory as well as the mayor of Gulf Breeze.

It's fairly easy to brush off this kind of UFO event as a fake. UFO sightings happen all the time, and, often, they're explainable. Sometimes it's an aircraft doing weird things or atmospheric phenomena such as "sun dogs" (bright lights called parhelia that appear to the left or right of the sun due to light refracted by atmospheric ice crystals).

But what about when they're recorded on video by Navy fighter pilots? Such is the case with the now-famous trio of recordings revealed in 2017 via news outlets and later officially released by the Pentagon. The videos' colloquial titles are "FLIR1," "GIMBAL," and "GOFAST."

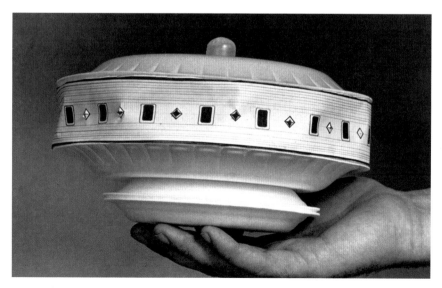

Ed Walters's paper plate UFO, as featured in the Pensacola News Journal

"FLIR1" was captured in 2004 when Commander David Fravor and Lieutenant Commander Jim Slaight of the USS *Nimitz* were flying over the Pacific Ocean during a training mission. Another Navy cruiser, the USS *Princeton*, asked them to investigate a strange object. Fravor reported upon what was later described as a Tic Tac–shaped object about 40 feet long hovering above the ocean at 80,000 feet. The object then plummeted toward the sea, where it lingered unmoving while the sea roiled beneath it. Another pilot sent out from the *Nimitz* to investigate had a forward-looking infrared camera (FLIR) and recorded the oddly moving pill-shaped object, though this second pilot never had a direct visual of the object.

In 2014 and 2015, strange objects off the Georgia and Florida coasts were seen spinning and advancing against a wind current over the Atlantic Ocean. The sightings happened on multiple occasions. The Navy pilots reported no obvious engines or engine exhaust but did note that the objects moved at hypersonic speeds. In the recording called "GIMBAL," the object at times looks like a black eggplant shape, but as it turns it appears more like a stunted moth. It spins as it flies forward in ways no known aircraft can, which has astonished aviators and engineers.

In "GOFAST," the white blip of an object is speeding along so quickly over the surface of the ocean that it takes several frustrating moments for the pilot to finally capture it in the center of the recorded visual field. When he does, he whoops, "Got it!" followed by triumphant laughter and then, "What the fuck is it?"

What the fuck is it, indeed.

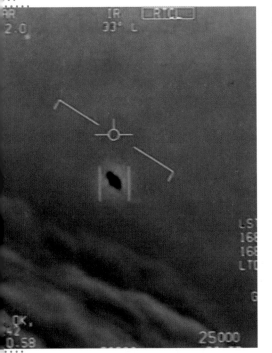

Unidentified anomalous phenomenon from the "GIMBAL" recording

Not If, But When

They are no longer called UFOs by serious journalists, scientists, and government officials. They are called UAPs, or unidentified anomalous (or aerial) phenomena. The term "UAP" seems less quacky, let's be honest, because "UFO" has become nearly synonymous with little green men, among other manner of aliens. UAP describes it exactly: stuff flying around that we can't identify. Especially when it's picked up by reliable sources (US Navy pilots, for example), and it's doing unexplainable things like flying at velocities with no obvious propulsion technology or at g-forces that would kill humans.

A 2021 Gallup poll reported that 41 percent of US respondents think that UAPs are made by alien life-forms. Not just that alien life exists, mind you, but that they've traveled here to our little greenish-blue marble. Astrobiologists have increasingly said that finding extraterrestrial life is plausible, though more likely in the realm of microbes than fully formed humanoids. In fact, there's no shortage of scientists who think it's only a matter of time (though maybe not in our lifetime) that we discover intelligent life not born on Earth. Still, there's quite a gap between what the public thinks and what might really be out in the great beyond.

The beliefs about extraterrestrial UAPs have been going on since long before the big alleged cover-up at Roswell, New Mexico, in 1947. In ancient history, celestial events, including comets and meteors, as well as optical phenomena such as the aforementioned "sun dogs" or lenticular clouds (white, saucer-shaped clouds of varying thickness), were usually explained by religious concepts, like angels or omens.

During World War II, Allied and Axis pilots saw "foo fighters," or balls of glowing light that never attacked the planes but were observed flying about in the air. Also called "ghost rockets" and "Russian hail," they were later believed to be St. Elmo's fire, a type of electrical weather phenomenon, or possibly something called ball lightning, a rare, unexplained sphere of glowing light that can be associated with thunderstorms. Some speculated that they were reflections off ice crystals in nearby clouds.

The first time "flying saucer" entered the lexicon was in 1947, when a civilian pilot claimed he saw several nonhuman craft flying in an angled line, like shining beads on a string, near Mt. Rainier. Kenneth Arnold, the pilot,

Atmospheric phenomena are often the cause of UAP.

claimed they were flying as fast as 1,700 mph (three times the speed of sound, and before humans had broken the sound barrier). He bobbled his airplane to make sure it wasn't something stuck to his window, then changed direction to see if the UFOs maintained their trajectory. They did. He reported his findings as soon as he landed, using terms like "saucer," "disk," and "pie-pan" to describe the objects' shape. A newspaper made the leap to naming them "flying saucers." Which, to be honest, is more impressive than "flying pie-pans." For the record, no definitive explanation was ever found for the sighting.

Other sightings happened around the same time. In the weeks that followed, there would be more than 800 UFO sightings of "flying saucers" or similar descriptions from the US and around the world. It was either a mass vacation for extraterrestrials or mass human hysteria. The same year, pieces of foil, tape, and rubber—part of a silvery weather or observation balloon—landed on a ranch near Roswell, New Mexico. The event would later turn into a cultural and pseudoscientific phenomenon that persists to this day and has become embedded in American folklore.

With Roswell, the UFO craze had officially begun. In the coming decades, more and more sightings would be reported. UFO saucer clubs popped up. Books on flying saucers began filling the shelves, with titles like *The Flying Saucers Are Real*. Even one of the founders of modern psychoanalytic theory, Carl Jung, added his *Flying Saucers: A Modern Myth of Things Seen in the Sky* to the pantheon. By the late 1940s, as the Cold War kicked off, fear of an arms race using extraterrestrial technology wasn't just something out of a Marvel movie. (Perhaps more worrisome was the possibility of advanced Soviet aircraft infringing on US airspace.) The US government launched Project Blue Book in 1952 to investigate UFOs, but it ended uneventfully in 1969 with the release of the Condon report, which concluded, "There was no evidence indicating that sightings categorized as 'unidentified' were extraterrestrial vehicles." Nevertheless, around 700 aerial phenomena went unexplained.

The academic world has pretty much held UFOlogy at arm's length for decades. It's hard to reproduce the data, there's a dearth of physical evidence, and there are countless other real-world explanations for the plethora of UFO sightings. Astronomer and astrobiologist Carl Sagan famously said, "The reliable cases are uninteresting, and the interesting cases are unreliable." Astrophysicist Neil deGrasse Tyson further warned, "The fact that you don't know what it is, is not evidence that you know what it is . . . the human sensory system sucks. That's why science exists. In science, eyewitness testimony is the lowest form of evidence there is."

So far, the list of things that were thought to be of alien origin but aren't include (deep breath) earthquake lights, drones, falling satellites, meteors, lens artifacts (including flares, lint, and bugs), weather balloons, spy balloons, birthday balloons, quadcopters, birds, airborne clutter like plastic bags, lenticular clouds, St. Elmo's fire, sun dogs, ball lightning, atmospheric sprites (electrical discharges in the mesosphere), distant aircraft, foreign

One version of what happened in Roswell, New Mexico

A stunning example of a sun dog, a type of parhelion in meteorology.

aircraft, parallax error (where things look like they're flying faster than reality because of an observational error), purposefully interfering radar spoofing, electronic warfare techniques, and experimental US aircraft. Oh, and, of course, swamp gas, which is UFOlogist slang for a conspiracy cover-up—the idea being that the government calls it "swamp gas" and hopes no one notices the truth.

If extraterrestrials have been visiting us, it does raise questions. Principally, if they're so advanced, why allow us to see them if they're trying not to be seen? Shouldn't they be more clever than that? Or have some sort of tech that allows them to go unnoticed? And *if* they can make it the many, many light years to visit us, why are there so many reported crashes once they finally get here? Finally, for those events we're allowed to see, why then hide the evidence (a huge government cover-up being the usual explanation)?

A big part of the popular narrative around UFO/UAP sightings includes a belief that governments might be hiding and reverse engineering any extraterrestrial tech they've found. If this is the case, we certainly don't seem to be making good use of it. As Michael Garrett, the chair of the SETI (Search for Extraterrestrial Intelligence) committee of the International Academy of Astronautics (a legitimate scientific institution) said, "Either they're not very good at reverse engineering, or there's nothing to be reverse engineered." It is certainly reasonable to imagine the former. If you handed Archimedes (who was born in 287 CE) an iPhone, he'd have no idea what was going on. But then Archimedes would at least tell other people—he often corresponded with mathematician friends in Alexandria about his discoveries.

None of this matters, however, because many UFOlogists claim there is an enormous underground organization covering it all up, which supposedly involved President Truman, Marilyn Monroe, and President Kennedy. Rumors say that Truman had a clandestine sit-down meeting with aliens, Kennedy was assassinated for wanting to reveal this, and Monroe was murdered for knowing too much. As for the alien tech, why haven't humans revealed it so we can use it to benefit humanity? Apparently, doing so would destabilize civilization as we know it. Carl Sagan noted on an episode of *Cosmos* regarding the possibility of being visited by extraterrestrial life, "Extraordinary claims require extraordinary evidence." But UFOlogists can comfortably say there *is* evidence—it's all just being hidden by powers greater than we know.

Which brings us to now. The *New York Times* published a shocking article in 2017 titled, "Glowing Auras and 'Black Money': The Pentagon's Mysterious U.F.O. Program." In it, the FLIR1 and GIMBAL videos were

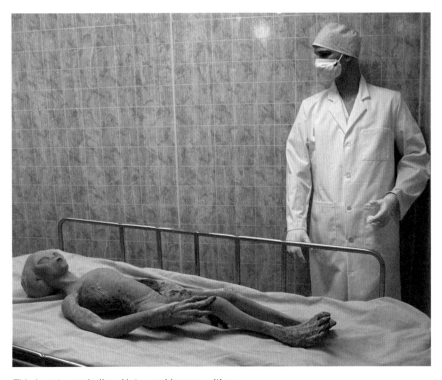

This is not a real alien. Not a real human, either.

revealed. The crux of the story revealed that beginning in 2007 an Advanced Aerospace Threat Identification Program did exist, and had received about $22 million in funding between 2008 through 2011, much of which went to Robert Bigelow, a civilian billionaire and aerospace entrepreneur. Though, to be clear, his billions originated from his Budget Suites hotel chain—a very terrestrial home away from home. Regarding the subject of finding extraterrestrials, Bigelow has said, "You don't have to go anywhere . . . It's just, like, right under people's noses." So ETs are among us, apparently. Bigelow also happens to be a friend of the late Senate majority leader, Harry Reid, who expressed interest in UFOs and helped push for the appropriation of funds for the project. But the program died in 2012—the results were apparently not dramatic or world-altering enough to warrant more funding.

The real consequence of the 2017 *Times* article was that suddenly talk of UAPs didn't have the whiff of pseudoscience anymore. Especially in the military, people were no longer as reluctant to speak about things they'd seen. In 2019, the Navy encouraged pilots to report UAP sightings without reprisals—and more reports came flooding in. Articles in reputable newspapers began to discuss the videos and what was being done (or not) in response to them. Then in 2020, the Senate Intelligence Committee mandated that the Director of National Intelligence work with the Secretary of Defense to concentrate on the gathered information around UAP sightings. The result: UAPs had finally reached a level of public legitimization. The Navy had been witnessing phenomena for years, after all.

Of the 366 new reports in 2022, 163 were balloons, 6 were clutter, and 26 were unmanned aircraft, such as drones. In October 2023, the annual report stated that "most UAP will likely resolve to ordinary phenomena." There is a very small percentage of UAP with "interesting signatures, such as high-speed travel and unknown morphologies [shapes]." Those will continue to be investigated. But luckily, no encounters with UAP have resulted in any adverse effects on people themselves. No alien abductions reported, in other words.

NASA is currently working with other scientists to better evaluate these UAP going forward, and they have plenty of work to do—they receive dozens of new reports per month.

The Government Comes Clean on Aliens

On June 25, 2021, the Office of the Director of National Intelligence released its declassified "Preliminary Assessment: Sightings of Unidentified Aerial Phenomena," with 144 UAP analyzed. Then, in 2022, a more comprehensive annual report was declassified and shared, with 366 further UAPs added in, and in 2023, there was yet another report with an additional 291 UAPs.

A few things to note about the reports: They used only high-quality reporting of UAP, which means that most of the sightings were made by the US military and commercial pilots, often recorded with multiple sensors. They also tended to cluster around training grounds or restricted airspace where the military were flying, which indicates a collection bias. The sources of the UAP were put into one of six categories:

- AIRBORNE CLUTTER, such as birds and plastic bags (Scott Kelly, a former astronaut and pilot, was once convinced he saw a UFO while flying off the coast of Virginia Beach. The radar intercept officer in the back of his Tomcat flighter jet was sure he saw a UFO. They turned around for a closer look, only to be faced with Bart Simpson—that is, a balloon of him. This is not in the government report, by the way.)

- NATURAL ATMOSPHERIC PHENOMENON, such as ice crystals or temperature fluctuations that messed with infrared or radar systems

- US GOVERNMENT OR PRIVATE INDUSTRY DEVELOPMENTAL PROGRAMS. The 2021 report was unable to confirm that these were responsible for any cases, which could mean they were . . . or they weren't.

- FOREIGN ADVERSARY SYSTEMS, or tech deployed by Russia or China, for example

- GAPS IN DOMAIN AWARENESS, which is a fancy way of saying that there is equipment error, misidentification, misperception, or sensor artifacts, like IR flares and optical effects like parallax

- OTHER. This is probably where the ET saucers reside, though the Office of the Directors of National Intelligence says: "We would group such objects in this category pending scientific advances that allowed us to better understand them." In other words, "We don't know what the fuck it is."

Is There Anybody Out There?

Is it possible that any non-Earth-originating life exists beyond our planet, including tiny organisms like microbes? It's quite plausible. Is it also possible that advanced, intelligent life is out there? Well, maybe. We'll leave that to the many scientists working on it, including those within the Galileo Project, launched by Harvard astrophysicist Avi Loeb, a controversial figure who hopes to use telescope technology and deep ocean retrieval of meteors to find evidence of extraterrestrial life. So far, the nonprofit SETI Institute, which looks for proof of life beyond Earth, has yet to find any. As for alien activity on Earth, we still haven't got any hard data there either.

We're listening but not hearing anything. Yet.

It's fairly common to hear that *they* have been to Earth—and that we'd all know about it if it weren't for the government concealing the truth. Remember the 2017 *New York Times* article that finally made it seem like it was okay to talk about UAPs in the light of day without being laughed at? Two of the journalists who wrote it, Ralph Blumenthal and Leslie Kean, published a shocking 2023 article on the website *Debrief* reporting on an intelligence officer turned whistleblower in 2023. He claimed that the US government has been allegedly hiding "intact and partially intact vehicles," aka UAP, for decades. He also claimed that analysis of the UAP fragments were "of exotic origin (non-human intelligence, whether extraterrestrial or unknown origin) based on the vehicle morphologies and material science testing and the possession of unique atomic arrangements and radiological signatures."

The whistleblower, David Charles Grusch, worked with the government's UAP Task Force from 2019 through 2021. He's not only a former combat officer who served in Afghanistan but also worked at the National Geospatial Intelligence Agency and the National Reconnaissance Office. He's already testified to Congress about these claims and stated that there has been a long-standing ongoing "Cold War for recovered and exploited physical material . . . to garner asymmetric national defense advantages." Real-life Hollywood-level stuff. He was reportedly told these things in confidence by high-ranking officials, and that the obfuscation was to prevent Congress from trying to oversee the UAP program.

Beyond what he said in the *Debrief* article, Grusch has gone on record in other venues that the craft may not necessarily be of extraterrestrial origin. It could be from "a higher dimensional physical space that might be colocated right here." In other words, a multiverse. Also, some of the craft are "football field kind of size."

On July 26, 2023, three ex-military veterans, including Grusch, testified before Congress about direct and indirect observations around UAPs. These included a US government cover-up of UAP crash sites, football field–sized UAPs, and ones repeatedly seen that look like dark gray cubes within a clear sphere. On the subject of other items at these crash sites, Grusch had an eye-popping exchange with South Carolina congresswoman Nancy Mace.

Rep. Mace: "If you believe we have crashed craft, as stated earlier, do we have the bodies of the pilots who piloted this craft?"

Grusch: "As I've stated publicly already . . . biologics came with some of these recoveries. Yeah."

Rep. Mace: "Were they, I guess, human or nonhuman biologics?"

Grusch: "Nonhuman, and that was the assessment of people, uh, with direct knowledge of the program that I talked to."

David Grusch testifies before Congress, July 2023

We are left with a tantalizing lack of information. Are there bits of tissue or cells, or whole corpses? In a subsequent interview with Joe Rogan, the word "variety" was thrown around with respect to the "entities." A variety of species, or forms of a species? It's unclear. And who actually saw them, and the downed aircraft? Grusch claims that multiple people have been harmed (he skirted around the question of murder) to keep the alien secrets secret. Where are the details of people harmed for knowing too much? These things were not disclosed during the public hearing or in further interviews.

Apparently, the activities of these life-forms on Earth are allegedly not all benevolent. Per Grusch in a June 2023 *News Nation* article, "The logical fallacy there is because they're advanced, they're kind . . . But I think what appears to be malevolent activity has happened."

But the most skepticism around all this incredible news comes from what isn't there: proof. Grusch has not seen any alien bodies or craft in person. He's repeating hearsay. And without physical proof or a true firsthand account, we're left with a story that sounds vaguely like a game of telephone.

The revelations from the congressional hearing hit the news cycle but were buried between the untimely death of Sinéad O'Connor and a revealing video about pajama shark mating behavior. It wasn't even headline news everywhere.

Why? Perhaps because we didn't learn much more than what we already knew. There are UAPs, we can't explain all of them, the public still has no direct evidence aside from grainy pilot videos and lots of hearsay, and, unfortunately, Grusch lacks any direct evidence of UAPs from an extraterrestrial source. If people within the US government know more, they're not talking. More data is what's needed.

In September 2023, NASA's UAP Independent Study Team—with its 16 experts in the military, aerospace safety, commercial innovation, tech, data, and space exploration—stated that "the threat to U.S. airspace safety posed by UAP is self-evident" and that "the detection of UAP is often serendipitous, captured by sensors that were not designed or calibrated for this purpose." And most pointedly, "To date, in the peer-reviewed scientific literature, there is no conclusive evidence suggesting an extraterrestrial origin for UAP."

Perhaps there is a big government cover-up. But the lack of evidence still means that UAPs and their origins will continue to live in the realm of not-yet-100-percent verified. One thing is for sure: If there are ETs, they're not nearly as clandestine as they could be, and yet the world keeps turning with all of us on it. Perhaps they'll be peaceful when they do make themselves known. The other option might be that alien life-forms decimate our planet or enslave every creature on Earth, which doesn't sound like fun. Having a UFOlogist saying "I told you so!" won't make it feel much better.

So until the next close encounter brings us irrefutable evidence, or the UAP investigations get better-quality data and better methods to evaluate UAPs, or the ETs actually show up and decimate us, we'll have to simply keep wondering if there really is someone or something else out there.

The Bermuda Triangle

On a clear December day in 1945, five Avenger torpedo bomber planes departed Fort Lauderdale, Florida, on a training run. Headed by flight leader Lieutenant Charles C. Taylor, the bombers intended to complete a routine flight pattern called "Navigation Problem No. 1." They would head due east from Fort Lauderdale, drop some bombs in a shallow coral reef, then proceed north over Grand Bahama Island before returning to Florida.

Taylor was an experienced pilot, with more than 2,500 hours of flight time under his belt. He had also completed a successful aviation tour in the Pacific Theater during World War II.

The weather forecast that day called for a few scattered showers. Nothing out of the ordinary. The pilots completed the first part of their training run without incident, releasing their bombs before proceeding north toward the Bahamas. It was then, however, that things took an abrupt turn for the worse.

Taylor radioed the flight tower in Fort Lauderdale.

"Cannot see land... We seem to be off course."

"What is your position?" responded the attendant at the radio tower.

"We cannot be sure where we are," said the increasingly distraught Taylor. "Repeat: cannot see land."

Contact with the planes was subsequently lost for 10 minutes.

The next communication came in from an unidentified pilot.

"We can't find west. Everything is wrong. We can't be sure of any direction. Everything looks strange, even the ocean."

After a tense 20 minutes without any communication, another transmission was intercepted by the radio tower:

"We can't tell where we are . . . everything is . . . can't make out anything. We think we may be about 225 miles northeast of base..."

Garbled communication followed before a final transmission was received in Fort Lauderdale, "It looks like we are entering white water . . . We're completely lost."

And that was the last anyone ever heard from Taylor or any of the other pilots on the training run that has become known as "Flight 19."

The Navy immediately launched a rescue mission, but, in a bizarre and unfortunate development, a Martin PBM

Avenger torpedo bombers flying in admirable symmetry

Mariner flying boat with 13 crew members sent to hunt for the Avengers also vanished.

In total, six planes and 27 men disappeared without a trace on that December day in 1945. For five long days afterward, the Navy, supported by the Coast Guard, searched a 25,000-square-mile radius of the Atlantic Ocean for any sign of survivors or debris from the missing planes. Nothing was ever found. Not even the smallest piece of wreckage.

Without a Trace

Draw a triangle that connects Bermuda, Puerto Rico, and the southern tip of Florida, and you've got the rough outline of the Bermuda Triangle. Also called "the Devil's Triangle," each side of the triangle is less than 1,000 miles. Within this area of high shipping and aviation traffic, a purportedly large number of ships and planes have disappeared without a trace over the decades.

Records of the number of missing ships or aircraft in the Bermuda Triangle differ by source and by the varying boundary lines for what is ultimately an invented area. Charles Berlitz, author of *The Bermuda Triangle*, one of the seminal tomes on the mystery, estimated that more than 100 planes and ships had vanished between 1945 and the book's publication in 1974, with some 1,000 people lost at sea. More recently, Gian J. Quasar, a "Triangle researcher," estimated an average of four aircraft and 20 yachts vanished yearly in his 2005 book *Into the Bermuda Triangle*.

To provide some context, on average, approximately 146 ships are lost in the world's oceans each year. When evaluating the world's most dangerous shipping areas, the World Wildlife Fund did not even include the Bermuda Triangle in a 2013 report. Indeed, the much deadlier "Coral Triangle," near southeast Asia, saw 293 shipping accidents between 1999 and 2013 alone, mostly the result of increased shipping traffic and poor governmental oversight. So the numbers of lost ships in the Bermuda Triangle

The HMS Atalanta. Totally inexperienced crew not pictured.

are not, on their own, particularly impressive. What intrigues Triangle researchers, however, are the ways in which the ships and airplanes have disappeared, often with little or no evidence left behind.

On November 6, 1840, the *Times* of London reported that the crew of a large French ship called the *Rosalie* vanished near Nassau in the Bahamas. "The greater part of her sails were set, and she did not appear to have sustained any damage . . . The captain's papers were all secure in their proper place . . . The only living beings found on board were a cat, some fowls, and several canaries half dead with hunger. The cabins of the officers and passengers were very elegantly furnished, and everything indicated that they had been only recently deserted."

In 1880, the British ship *Atalanta* set sail from Bermuda and headed to England, only to disappear en route, along with all 290 hands on board, the vast majority of whom were inexperienced cadets on their first training voyage.

In 1909, Joshua Slocum, the first man to sail solo around the world, disappeared in the Bermuda Triangle in his yawl, a two-masted sailing boat

The USS Cyclops *in better days*

called the *Spray*. Slocum was a very experienced sailor who had survived terrible storms at sea in his voyage around the world, yet he appears to have vanished on what should have been a routine sailing trip from Martha's Vineyard to South America.

In 1918, the 542-foot ship the USS *Cyclops* departed Barbados with a crew of 309 men only to disappear without a trace before it could arrive in Norfolk, Virginia. A massive search was conducted, but no sign of the ship was ever found, and the US Navy stated, "The disappearance of this ship has been one of the most baffling mysteries in the annals of the Navy, all attempts to locate her having proved unsuccessful. Many theories have been advanced, but none satisfactorily accounts for her disappearance." Even President Woodrow Wilson purportedly said, "Only God and the sea know what happened to that great ship."

In 1963, the *Marine Sulphur Queen* embarked from Beaumont, Texas, to Norfolk, Virginia, with a full cargo of molten sulfur and a crew of 39. On the morning of February 4, the ship sent a routine radio call from a position

270 miles off of Key West. It was the last anyone would hear from them. Attempts to radio the ship after February 4 were unsuccessful and the ship failed to arrive as expected in Norfolk. An extensive search was conducted, recovering a few life jackets and some floating debris, but nothing more. According to the legend of the ship, no official explanation was ever offered.

And the list goes on and on.

But what's really happening here? These disappearances are undeniably puzzling and have certainly attracted a great deal of speculation over the years, including some bold forays into the realm of pseudoscience. The first mention of the Bermuda Triangle as a distinct area occurred in a *Miami Herald* article by Edward Van Winkle Jones on September 17, 1950. From there the idea was popularized by Vincent Gaddis in a 1964 article in *Argosy*, a widely read mid-century magazine with circulation close to 1 million, followed by his book on mysteries of the sea, *Invisible Horizons* (1965). Then, in the late 1960s and early 1970s, paranormal ideas of a "triangle of death," where ships and planes mysteriously disappeared, really took off. Several pulpy publications about the "Devil's Triangle" and the "Limbo of the Lost" became bestsellers, particularly Charles Berlitz's 1974 book *The Bermuda Triangle*.

If Not a Time Warp, then Aliens . . . or Atlantis

Explanations of this supposed phenomenon were as wide and varied as the Atlantic Ocean itself. One oft-repeated explanation involves a space and time warp that appears occasionally, without warning, at various points in the Triangle. This theory is buttressed by the atmospheric aberrations reported by eyewitnesses, such as those dramatically delivered in the radio transmissions from the five Avengers planes in Flight 19. The unfortunate ships and aircraft caught in such a time warp become trapped in another dimension, where they await their uncertain fate: Either they will be freed from the time warp at some future date and return to their home ports without having aged at all . . . or their ships will one day be found floating on a lifeless sea with only skeletons aboard.

Then there are the UFOs. This theory was popularized by John Wallace Spencer in his book *Limbo of the Lost*. Spencer became enthralled with the

UFO theory while investigating the disappearance of the US atomic submarine *Scorpion* (which was eventually found). He became convinced that the only way for ships and planes to disappear entirely was for aliens to abduct them.

"Since the complete disappearance of 575-ft vessels in calm seas 50 miles offshore or commercial airliners going in for a landing cannot happen according to earthly standards, yet *are* happening, I am forced to conclude that they are being taken away from our planet," he claimed.

If the threat of abduction from above wasn't enough, there was also the menace rising from below. In his book *Invisible Residents,* Ivan Sanderson suggested that the sea itself was inhabited by aliens who are

Underwater alien ship just waiting for its next victim in the Bermuda Triangle

largely uninterested in the humans occupying the world above them but will on occasion bother themselves to kidnap several of us for study. Which of course explains the ships found in the Triangle without anyone left onboard.

But wait, wait—there could be something even worse afoot than alien abduction: alien energy signals. The idea here is that a signal device was left in the Bermuda Triangle by exploratory aliens in ancient times. Apparently, the device was installed to send signals into space that could direct alien explorers to the best landing path on Earth. (Why an alien spaceship would want to land in one of the least navigable sections of the planet sort of begs explanation, but anyway.) Tragically, when the alien device sends its occasional signals into outer space, its powerful alien beams impact not only the navigational instruments of ships and aircraft in the region but the human mind itself! And even worse, the beam may be so powerful it can destroy any vessel directly in its path.

Edgar Cayce thinking about Atlantean crystal-based technology

Another theory was "realized" by mid-twentieth-century psychic and clairvoyant Edgar Cayce, who surmised that the lost city of Atlantis sank off the Bahamas, but that its powerful energy source technology continues to pulsate away, exerting an undue influence on the Bermuda Triangle region. Regrettably for the Atlanteans, their sophisticated crystal-based technology was not powerful enough to prevent their entire civilization from disappearing into the Atlantic Ocean—yet it remains powerful enough to mess with twenty-first-century navigational equipment.

Time warps. UFOs from space. Aliens from deep ocean depths. Extraterrestrial space beams. Atlantean laser technology. With such an array of forces and obstacles stacked against it, no wonder that a journey within the Bermuda Triangle is such a difficult thing to survive.

The Devil's Sea

The Bermuda Triangle is one of the places on the planet with both heavy shipping traffic and an occasional agonic line, where a compass points to true north rather than magnetic north. A similar area, off the coast of Japan, is known as "The Devil's Sea." Unsurprisingly, the Devil's Sea is also known for an unusually high number of strange disappearances over the years. According to Charles Berlitz in his book *The Bermuda Triangle*, at least nine ships disappeared in the Devil's Sea between 1950 and 1954. Similar to Bermuda Triangle incidents, those in the Devil's Sea often involve large freighters—with theoretically powerful engines and radio communication systems—that disappear abruptly and with little or no signs of debris.

As with the Bermuda Triangle, natural explanations can be offered. In addition to the significant navigational problems caused by compass variation in the Devil's Sea, there are numerous underwater volcanoes and vast fields of methane hydrates that can abruptly explode with methane gas. If a ship was unlucky enough to be near the explosion of an underwater volcano or a methane hydrate vent, its buoyancy would be severely disrupted and it could sink immediately, with no time to send a distress signal. Finally, with all the underwater seismic activity, the area is frequented by tsunamis.

Once again, environmental variations, not paranormal phenomena, are the culprit behind the disappearances—no matter how creatively named the region.

An Ill Wind That Blows No Good

To be fair, the Bermuda Triangle is indeed a strange place, but not on account of any supernatural forces. A variety of unusual environmental factors make the Triangle area particularly difficult to navigate for ships and aircraft alike.

An agonic line has at times run through the Bermuda Triangle, including in the early twentieth century. An agonic line occurs where true north and magnetic north are in alignment, but the line can change over time depending on variations in the Earth's magnetic field. While that may seem like a subtle navigational change, the impact can be substantial. The difference between true and magnetic north is known as "magnetic declination." And that amount of variation can change by as much as 20 degrees, so it must be constantly accounted for by navigators. True north is a fixed point on the Earth's surface: the North Pole. Magnetic north, by contrast, is the direction a compass needle points to when it aligns with the Earth's magnetic field. However, the catch is that the Earth's magnetic field changes over time and space, making magnetic declination crucial for navigation. When true and magnetic north are in alignment, accounting for compass variation is unnecessary. If this alignment is missed, a mistake even an experienced navigator could make, the course could quickly veer off by several degrees—and, therefore, hundreds of miles.

Then there is the Gulf Stream to consider. It's an extraordinarily fast and turbulent ocean current that delivers massive amounts of warm water from the Gulf of Mexico into the Atlantic Ocean. And it flows directly through the Bermuda Triangle area, quickly erasing and scattering debris and signs of disaster, accounting for the "mysterious disappearance" of so many of the ships and aircraft lost there.

The interplay of Caribbean and Atlantic weather patterns also impacts events in the Bermuda Triangle (hurricanes, anyone?). The clash between warm and cold air currents can lead to rapidly developing weather incidents on a local scale. Sudden thunderstorms and waterspouts can spring up, seemingly out of nowhere, quickly overtaking ships and aircraft.

The diversity of ocean floor topography in the Bermuda Triangle also plays a role, varying between extensive shoals around the islands to some of the deepest marine trenches in the world. The ocean floor is further in a constant state of flux from the strong currents, creating new and unexpected navigational hazards that also contribute to disasters.

The considerable environmental factors that make the Bermuda Triangle such a tricky place to navigate can be easily compounded by human error or malfeasance. Piracy is another factor in the area, historically anyway. The relative emptiness of the region's waters makes it a convenient place to stage a disappearance or commit a murder. Additionally, the Bermuda Triangle is regularly frequented by small boats and pleasure craft traveling between Florida and the Bahamas. According to the US Coast Guard, many of those crossings are attempted on boats ill-suited for the journey, or crewed by people with insufficient knowledge of the many hazards of the area.

The US Coast Guard's official statement about the Bermuda Triangle is short and to the point:

"The Coast Guard does not recognize the existence of the so-called Bermuda Triangle as a geographic area of specific hazard to ships or planes. In a review of many aircraft and vessel losses in the area over the years, there has been nothing discovered that would indicate that casualties were the result of anything other than physical causes. No extraordinary factors have ever been identified."

So if the Coast Guard, which has spent countless hours patrolling the Bermuda Triangle for decades, is not impressed, neither are we. After investigation, many of the seemingly mysterious disappearances in the Bermuda Triangle turned out to be more tragic and less mysterious than originally thought. Larry Kusche, after extensive primary source research, debunked many of the legendary Bermuda Triangle disappearances in his book *The Bermuda Triangle Mystery—Solved*. Here's a bit more insight into some of the most famous Bermuda Triangle disappearances, including those recounted above:

- **Flight 19** was a mix of controls malfunction and pilot navigational error. A perfect storm of events occurred: the failure of Taylor's compasses, poor radio reception, the approach of night by the time the flight was in trouble, the sudden arrival of bad weather, the military discipline that prevented some of the pilots from flying away to safer areas, even though it is clear from their radio transmissions that some knew they were off course. The rescue plane that was lost was also a type of plane known as a "flying gas tank." Something as small as a spark from an illicit cigarette could have easily caused an explosion.

- **The *Rosalie*** probably never even existed. French records contain no mention of the ship, and the entire legend is based on a single misunderstood (or fabricated) story from a Nassau correspondent for the *Times*.

- **The *Atalanta*** was crewed by an overwhelmingly inexperienced group of cadets and almost certainly sank in bad weather with a likely factor being not enough experienced sailors on board to save the ship. It probably wasn't even in the Bermuda Triangle when it sank, as it was heading for England.

- **Joshua Slocum** couldn't swim—if he fell overboard from the *Spray*, he was a goner.

- **The USS *Cyclops*** had serious structural problems that were later identified when two of its sister ships also disappeared during World War II.

- **The *Marine Sulphur Queen*** had . . . literally . . . molten sulfur as its cargo. That, combined with some structural weakening from the aging ship, almost certainly led to it breaking in half and sinking so fast no one could even release a distress call.

The ocean is vast, unpredictable, and exquisitely dangerous, qualities that are amplified in the Bermuda Triangle area because of the variety of environmental factors at play there. It's tragic that so many ships, planes, and lives have been lost in the region, but it only distracts from those tragedies when we focus on outlandish theories that include alien laser beams and Atlantean power sources. These were real people who met untimely ends and deserve to be treated with the kind of respect that comes with that. By focusing on improving navigational awareness, preparing for unlikely weather events, and advancing communication strategies, perhaps we can reduce the lives lost in the so-called Devil's Triangle.

Crop Circles

July 7, 1996. This was the day a crop circle appeared in Wiltshire, England, miraculously imprinted in a farmer's field near the famed neolithic Stonehenge. The edges of the design were sharp and crisp, cleanly pressed into the crops. From an airplane or helicopter, it was a breathtaking sight: a fractal pattern, known in the mathematical world as a Julia set, consisting of a curving circle spiraling outward with clusters of circles of varying diameters. This was no random signature of a whirlwind. The Julia set was precise, purposeful, exquisite, and clearly done by an intelligent being.

Of course, this was not the first crop circle the world had seen. In fact, at the time, many people brushed off crop circles as the works of mischievous humans, rather than aliens from outer space. People had demonstrated that they could be made by walking around with planks and ropes, pressing down corn and wheat in seemingly random fields across the world. But this new crop circle in Wiltshire was unlike the hoaxes of the 1970s and '80s that had been outed as human made. What happened near Stonehenge set itself apart, and timing had everything to do with why.

A pilot had flown above the field late that afternoon and noticed nothing amiss. Less than an hour later, on the return flight, the exquisite 600-foot fractal suddenly marked the land. A farmer working the fields had seen no one trespassing on his property that day. Guards at Stonehenge had spotted no hoaxers infiltrating the nearby fields. And then, one eyewitness came forward, known only (and mysteriously) as "M."

According to a 2022 article in the *Skeptical Inquirer* by Benjamin Radford regarding this "Stonehenge Surprise," M claimed they had left their car on the small highway that separated Stonehenge from the field, joining others who had stopped to view something strange on the farmer's land.

"There was an apparition, an isolated mist over it and as the circle was getting bigger the mist was rising above the circle. As the mist rose it got bigger and the corn circle got bigger," M said. "There was a mist about two to three feet off the ground and it was sort of spinning around and on the ground a circular shape was appearing, which seemed to get bigger and bigger as, simultaneously, the mist got bigger and bigger and swirled faster."

Surely, the Julia set could not have been made by a person (or people) in such a short period of time without someone noticing. So who—or *what*—created it?

Aliens and Fairies and Devils!

The creepy 2002 movie *Signs* may come to mind when you think of crop circles, but you might be surprised to know that humans have been noticing patterns in crop fields for hundreds of years.

Exhibit A is a 1678 news pamphlet entitled "The Mowing-Devil; or, Strange News out of Hartford-Shire," which tells the story of a farmer who was arguing with a worker over the cost to mow his field. He refused to pay a worker the asked price, stating "the Devil should Mow it, rather than He." That night the oat field appeared to be on fire, but by the next morning it had been

An elf circle, or pixie ring. Magic or mushrooms?

strangely mowed in a way "that no mortal man was able to do the like." An accompanying image showed a devilish creature cutting the oats in a circular fashion, offering a supernatural explanation for the strange circle.

For centuries, people have been entranced by other unusual circular land formations, such as fairy rings. Also called "elf circles" or "pixie rings," these patterns are often found in fields or in woodlands. They looked like an area marked by fantastical creatures for a place of revelry, and they have been the subject of European literature since the thirteenth century, often with warnings for those who might want to enter them. In the nineteenth century, English naturalist John Aubrey theorized that such mysterious patterns "are generated from the breathing out of a fertile subterraneous vapour," and earlier, in 1686, English naturalist Robert Plot proposed that the circular patterns descended from the sky. (In reality, they are created by certain types of mushrooms growing in a circular formation.)

But crop circles are quite different from fairy rings. One speculator in the late 1800s, amateur astronomer John Rand Capron, described a crop circle as: ". . . some prostrate stalks with their heads arranged pretty evenly in a direction forming a circle round the center . . . suggestive of some cyclonic wind action . . ." In the twentieth century, crop circles were noted as early as 1932, also in England, as well as in 1960, where a wheat field in Wiltshire was noted to have a "spiral flattening" thought to be made from a meteor strike. In fact, as Patrick Moore explained in a letter to the *New Scientist* in 1963, "it looks very much as though they, and the crater, were caused by something which came from the sky."

Rise of the Croppies

Cyclonic winds. Meteors. Vapors. All offered scientific explanations for crop circles, but in January 1966, an event in Australia added an extraterrestrial dimension to the crop circle phenomenon.

George Pedley was driving along the Tully River when he heard a strange hissing sound. He then saw "a spaceship rise at a great speed . . . about 25 yards in front of me." The blue-gray object, 25 feet across and 9 feet high, then spun up and away. Upon investigating, he found that there was a circular mat of reeds that appeared to have been flattened by the craft but had actually been uprooted and were floating in the lagoon like a round plate. A formal investigation pointed out that "willy willys" (whirlwinds) or a waterspout were the more likely of the reeds' displacement.

The alleged UFO sighting, and its connection to the formation of a "reed circle," struck a chord with the masses. The official investigation that followed did little to stymie the now-growing murmurs that UFOs could be the reason why crop circles began appearing with startling regularity in the 1970s. The designs had circles, lines, and triangles, sometimes with branched antennae-like patterns, sometimes with connecting lines. Often, the wheat or corn would be bent over and flattened, with the stalks intact and uncut. They often popped up near cultural heritage monuments with ancient histories, such as the Avebury neolithic stone circles and also in Wiltshire, England (which seemed to be a recurring hot spot for crop circles). Usually, they

A theoretical UFO leaves a crop circle behind with uncanny geometric precision.

appeared abruptly, often overnight. And they didn't show obvious signs of human involvement. There were no footprints, for example, no tracks leading to or from the fields. Some people noted strange sounds occurring nearby, or glimpses of UFOs.

By the 1980s, "crop circle" had entered the vernacular (including the *Oxford English Dictionary* in 1988). Crop-circle enthusiasts, or "cereologists," so called because crop circles happen in grain fields, or the more casual term, "croppies," believed they were not observing a man-made phenomenon. United by a silly name, they argued passionately about the causes of the intricate designs. Were they markers for extraterrestrial visits? Were they communications from Earth itself to stop ruining global ecosystems?

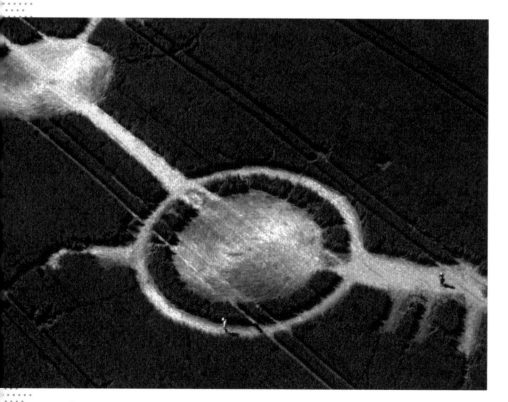

Crop circles appearing in Wiltshire, England, that cannot be explained away by intoxicated wallabies.

Some people stuck to weather phenomena as a potential source. Others indulged in ideas that were laughable yet seriously discussed, such as the report that Tasmanian wallabies, intoxicated from snacking on opium pods, were hopping around in dizzying patterns, creating crop circles in poppy fields. Of course, that didn't explain the large number of crop circles showing up in England, given their lack of opium fields . . . and even more notable lack of wallabies.

In the absence of an explanation for these strange formations, people's imaginations ran wild. Some believed ancient spiritual forces caused crop circles, while others guessed that bizarre weather oddities, international cover-ups, or a secret plot by MI6 were making it hard to find the real cause. And what about the fact that when standing in a crop circle, some people claimed they felt a visceral response? One woman said she felt like she was "home." Others experienced hands tingling when touching the crop circle, and they also reported that their hands changed colors.

Legitimate voices began to wonder if there was a natural origin. Waseda University professor of plasma physics Yoshi-Hiko Ohtsuki asserted that they were caused by "the plasma vortex," a phenomenon relating to a type of ball lightning and plasma fireballs. Using models, he theorized that magnetic and axial electric fields might explain the origin of crop circles. Stories from Japanese mythology of maidens swooping down from the sky almost matched the concept, if you blurred your imagination a little. Or a lot. Even the *Economist* reported that the mystery could be over. A former physics professor, Terence Meaden, believed that plasma vortices caused a mirage-like light easily mistaken for a UFO. It supposedly made sense that crop circles happened in places near Stonehenge and the Avebury circle, where neolithic peoples might have witnessed similar light displays.

Some croppies pointed to the Gaia theory expounded by James Lovelock in the early 1970s. Named after the Greek goddess of Earth, the theory posits that Earth is a self-controlled regulatory system (much like a single organism), and by extension, the crop circles are a warning signal to humans that they are doing too much damage to the environment. (Apparently, the best that Mother Earth could do to alert us to her distress was create cryptic patterns in our crops.) Pollution and melting glaciers aren't nearly as cool of a distress signal, it seems.

Crop circle caused by, possibly, MI6, or plasma vortices, or mad Earth

But UFOlogists didn't think much of these alternative theories. If crop circles were messages, then the messages were getting more complicated. The intricacies of crop circles began to increase in the 1980s and '90s. Were these just areas where UFOs had landed, or truly inscribed messages? Books were being made and sold discussing the different theories of croppies. Crop circle tours were making money. Farms were being trampled by tourists looking for the best, or most elaborate, designs. Even the rock band Led Zeppelin featured a particularly cryptic-looking one in Eastfield, Alton Barnes, in (surprise!) Wiltshire, on their boxed set of albums released in 1990. By then, the crop circle craze had reached unfathomable heights.

The Hoaxers Come Clean

The beliefs of die-hard croppies were strongly challenged in 1991 when two Southampton men, Doug Bower and Dave Chorley, finally raised their hands to admit that they'd created the crop circles in southern England that had led to the modern crop circle pandemonium.

Bower and Chorley had heard of the reported UFO sightings and reed circles in Australia's Tully River, where they got the idea. They used a four-foot plank of wood, rope, a ball of string, and a baseball hat with a sight-wire on the brim to produce their artwork. Step by painstaking step, they used the plank to press down the wheat stalks while holding on to the string from the center of the circle in order to create a perfect shape. All told, their work took a couple of hours. They could enter a field after nightfall and quickly get to work. By the time dawn came, they were long gone.

Starting in 1978, Bower and Chorley produced about 25 to 30 crop circles in each farming season, keeping a careful log of their work. This was the

Doug Bower and Dave Chorley, the not-so-extra terrestrials responsible for driving croppies wild

reason why crop circles tended to occur in southern England, near where the pranksters lived. It also explained why the creations were close to roads, and done only at night. Bower and Chorley admitted to the hoax once there was talk of a government-funded investigation. A joke on the world was one thing; wasting taxpayer money was another.

By 1991, when they finally confessed, the stealthy pair had made over 200 crop circles, and their work had inspired copycats. After their methods were revealed, crop circles became a global phenomenon, and they still appear worldwide to this day. Some crop circle creators are pranksters, some are artists, and some have their own silent reasons. Some proclaim their fandoms (there are Star Wars–themed crop circles around). Some use barstools to enter the fields, hopping from stool to stool, so as not to leave footprints. GPS is now a useful tool, especially for intricate designs that also include advertisements paid for by HBO, Papa Johns pizza, and the makers of *South Park*, with a giant Eric Cartman and his obnoxious signature stare coming at you from a golden field.

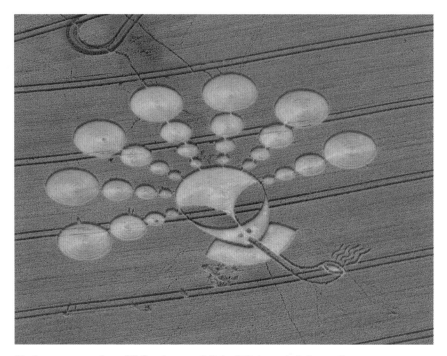

Unclear message from ET. Smoke more? Relax? Get new hairdresser?

But, given all this, do people still believe that crop circles are not made by humans? Absolutely. Given that not every crop circle creator has de-anonymized themselves, speculation continues. In the corners of the internet and the world, you can still find chatter about magnetic fields, physiological effects on humans, strange noises and frequencies nearby, how much time it truly takes to make them, and arguments about "real" versus "hoax" crop circles.

So what about that 1996 Julia set in England that supposedly appeared in less than an hour, in late daylight? And what of the eyewitness account from the mysterious M?

It turns out that M's account was spread through word of mouth, the source of a source, and the real M has yet to step forth and give a statement about what they saw. Furthermore, there is no other evidence that clusters of motorists were watching the so-called UFO-like event. The guards at

Stonehenge likely weren't keeping a close eye on the field (they aren't paid to guard the neighboring farmer's field), which can't even be seen very well from the road, given that the land doesn't slope downward to the Julia set. The farmer didn't wade deep into his field to prove that no one was in it at the time the crop circle was created. As for the pilot who claims it was made in less than an hour in daylight? Allegedly, there was a doctor in the airplane who took photos. That doctor, and those photos, have never surfaced.

If the gold standard of proving phenomena is a perfect eyewitness account, then the platinum standard might be an actual confession. In 1996, a man named Rod Dickinson told researcher Michael Lindemann that the reason why it seemed supernatural for the Julia set to be made in 45 minutes was because it had actually taken 2 hours and 45 minutes. Not on July 7 in the daytime, when it was first seen, but in the dark hours before dawn.

"It was made the previous night," Dickinson said, "by three people." He went on to explain how the process unfolded. "You start with the large central circle, which is placed right next to a tram line. People asked why it had the large central circle, which is a little out of place in a Julia Set. Simple. To avoid damaging surrounding crop, you have to have a large central area already layed [sic] down, from which you can measure out diameters to other parts of the formation. After making the first circle, they measured out a work line for the rest of the formation."

The magical, mathematical, surreal Julia set was as ordinary as other crop circle hoaxes come.

· · · · ·

It's funny how geographical oddities can seem inexplicable at first, but the dearth of information surrounding their creation is also fodder for supporting the ensuing conspiracy theories. The Julia set was crafted by some very clever—and very quick—humans. But then again, croppies and cereologists will say that the confessions of human pranksters equally allow for the possibility that extraterrestrials, or some yet unidentified force, created "real" crop circles.

And so crop circles will no doubt continue to show up for years to come, so long as they can generate just a little bit of speculation, and a lot of wonder. That, or whenever someone wants to proclaim their love for Star Wars.

Ghosts and Ghost Hunting

Specters. Wraiths. Phantoms. Demons. Spirits. Poltergeists. Apparitions. Wights. Revenants. Bogeymen. Elementals.

Ghosts.

Their stories fill us with dread and exhilaration. An entire autumn holiday revolves around the historical practice of keeping them at bay with hollowed-out pumpkins and gourds. They can be vengeful, sad, unfulfilled, playful, troublesome, helpful, protective, deadly. Ghosts show up in cultures and religions throughout the globe, from Mesopotamia and ancient Egypt to references in the Old Testament in the Bible, in which the ghost of Samuel is summoned by the witch of Endor.

"What do you see?"

The woman said, "I see a ghostly figure coming up out of the earth."

"What does he look like?" he asked.

"An old man wearing a robe is coming up," she said.

Then Saul knew it was Samuel, and he bowed down and prostrated himself with his face to the ground.

Buddhist and Taoist communities throw a Ghost Festival. In many parts of Asia, the Hungry Ghost Festival is celebrated. Ghosts are called bhoots in South Asia and gwisin in Korea. The Luo people of Kenya believe that the spirit (tipo) continues on after death and can visit families in visions or make babies cry. In Scandinavian folklore, they are called gjenganger, and in Mexico, ghosts (fantasmas) are extensively incorporated into the culture.

In short, ghosts are everywhere.

Lanterns glow on the Zijiang River for the the Zhongyuan Ghost Festival

Just ask your friends. A casual survey will reveal a handful or more of people who have witnessed them in person. Or who know someone who's seen one.

In an informal polling of friends and acquaintances, we heard stories including: "Sometimes I see a small child in my home wearing old-fashioned clothes, out of the corner of my eye" and "Sounds like the man in black I once found sitting at the family home's dining table one late night around 1992-ish" and "The dog keeps barking at an empty corner of the house and I know it's the dead person who used to live here."

These reports came as no surprise. A 2019 Ipsos poll showed that 46 percent of Americans believe in ghosts. A YouGov America poll in 2021 put the number at 41 percent. (The latter poll also showed that 43 percent of Americans believe in demons, but that's for another discussion.) Other polls often show numbers north of the 50 percent mark. That translates to at least 165 million people in the US believing in ghosts. Let that sink in for a moment.

The supernatural and paranormal have always been part and parcel of human existence. There are stories—parts of the mythologic fabric of societies around the world—and also real-life experiences that frighten and confound their observers. And there are plenty of people who are willing to try to prove they're real. Or, at least, who are hoping to exploit people's beliefs about ghosts for their own end.

A Brief History of Ghost Busting

Why do we try so hard to prove that these specters are real? Before ghosts were avidly hunted with machines and cameras at the ready, there was a movement called Spiritualism, a Western infatuation with ghosts and the undead that rose to prominence during the nineteenth century. Some say Spiritualism began in 1848 in Hydesville, New York, when Maggie and Kate Fox became famous for communicating with "ghosts" via knocking noises on the walls or furniture. Mediums all over the country jumped on the fad, selling their abilities to commune with the dead. After the US Civil War, when 2 percent of the population (or more than 700,000 people) died, families longed to connect with their loved ones after suffering massive loss, which created a wealth of desperate customers. Even Mary Todd Lincoln attempted to speak to her dead son, Willie, via a séance held at the White House.

It was almost too easy for mediums to trick heartbroken family members into believing that their loved ones were still around. At the time, electricity was a new but increasingly common phenomenon that seemed almost magical. Why not believe we could also reach out to the energy of the dead? "Spirit photography" took advantage of double exposed film to show shadowy figures hovering in portraits of clients hoping their departed loved ones hadn't really left them. Tables tilted via hidden hands. Hidden pipes could introduce strange sounds and voices in séances, along with strings and pulleys that made objects move or provide knocks from the dead.

The infamous Kate and Maggie Fox

Mary Todd Lincoln and the ghostly image of her beloved Abraham

The famed inventor Thomas Edison was just as hopeful as others about an earthly connection to the spiritual realm—he tried to create a "spirit telephone" to help. Even after the notorious Fox sisters admitted they were frauds, the hope that ghosts were real remained a fixture of public fascination. In one particularly famous case replete with splashy headlines, a spirit photographer from Boston named William Mumler was brought to trial in 1869. He was then found guilty of fraud for his double-exposure "spirit" photos. In an ironic convergence of fraudsters, it was P. T. Barnum who testified against Mumler. Barnum argued

A ghost haunts the living by making an utter mess of things, circa 1865

for truth against the fakery, which is remarkable given that Barnum himself benefited from all manner of fakery, including his cobbled-together Feejee mermaid.

Fast forward to the end of the twentieth century, when the *Ghostbusters* films introduced an entirely new concept to ghostly pursuit. Here were a bunch of ordinary but academically minded guys who brought science to the investigation of ghosts. They used handheld psychokinetic energy meters, proton packs to lasso misbehaving ghosts, and portable ghost traps. The ghostbusters weren't simply investigating the phenomena but understanding and controlling them as well. Instead of wondering whether ghosts were real or not, or just flat-out running away screaming from phantoms,

Peter Venkman, Ray Stantz, and Egon Spengler prepare to be slimed by ghosts in a very scientific manner.

the ghostbusters offered the (fictional) possibility that you could proactively do something about ghosts. You could actually engage with the spirits in these films, even capture them and remove them from a haunted location. Suddenly, regular people could conceivably get to know spirits on a scientific level.

Yes, it was a movie (with a giant angry marshmallow man in the original), but it's hard to ignore its influence on modern-day ghost hunters. Ghost hunting became immensely popular in the decades following the movie releases. And like the characters, paranormal investigators now also possess fringe scientific equipment used by those who are shunned by mainstream academia. There are real people like the *Ghostbusters* character

Winston Zeddemore, the ex-military regular Joe, who enter haunted areas, devices first, looking for the truth. And they're heroes (at least in their own minds). It's easy to see the appeal.

As the twenty-first century kicked off, paranormal television started collecting lots of viewers. *Ghost Hunters* was wildly popular beginning in 2004 on the Syfy channel and continuing for more than 16 years across several networks. Early on, it was followed by *Ghost Adventures*, *Paranormal State*, *Kindred Spirits*, and *Ghost Nation*. On these shows, viewers were introduced to a variety of techniques used to detect paranormal activity. Jump scares were an expected part of the action, including EMF (electromagnetic field) meter needles bouncing, puffs of cold air hitting the hunters, and figures or orbs of light appearing in dark corners. Of course, it was all accompanied by yelling, screaming, scrambling, and chaotic videos, culminating in a desired spine-tingling effect.

The data-gathering equipment of modern-day ghost hunters, both professional and amateur, includes video cameras, night-vision equipment, audio recorders, digital thermometers, and other electronic regalia, such as EMF meters and infrared motion detectors. Photos are featured items that demonstrate supposed proof of ghosts. Fuzzy smudges where there shouldn't be any are a common theme. Face shapes that inexplicably show up in shadows are another. Videos are yet another way of capturing not just a snapshot, but the ghost in action. Temperature is further thought to be a classic sign of a ghostly presence, manifesting as sudden unexplained drops in degrees.

Dowsing rods (see page 275), carbon monoxide monitors, infrasound equipment, Ouija boards, and trigger objects (things like wedding rings or other objects of emotional importance) are often used in attempts to bait and catch ghostly presences. Compasses and Geiger counters (used to detect gamma radiation and X-rays collectively known as photons) are employed to search for fluctuations in magnetic fields or radiation. And then there is the Ovilus, which transforms "environmental readings" into words. The robotic, nearly incomprehensible voice will say things like "magic" or "mustard" when it picks up on . . . *something*. The description on several ghost-hunting equipment sales websites does not explain what the "environmental readings" are, exactly, or what their accuracy is like (or on what basis you can judge accuracy for a device like this). The words, however, are promised not to be randomly generated.

Okay...

But of all the equipment used, EMF meters appear to be the gold standard. They often possess a brightly colored monitor, which picks up static and moving electrical charges that create a magnetic and electric field, something that is usually emitted by sources created by people, such as electrical lines or appliances. Ghost hunters believe that somehow the souls of the dead transferred their life energy (or "mind energy") into an afterlife energy that creates perturbations in the EMF fields—any unusual readings on an EMF meter are considered proof that there are ghosts nearby.

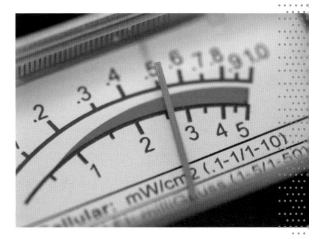

EMF meter indicating a nearby ghost. Or a nearby hair dryer.

But what were the ghost hunters on TV and off-screen really finding?

The answer is a whole lot of . . . nothing. Ghost hunting is an endeavor whose lifeblood exists in the world of fuzzy voices, blurry photos, temperature fluctuations in drafty places, and EMF needles that bounce around randomly. In other words—false positives. The EMF meters, which do have a practical usage in measuring EMF in appliances, are commonly employed by ghost hunters to detect a particularly long wavelength of about 50Hz to 20kHz. Why these parameters? It appears arbitrary. Which means they might pick up on, say, a lightning strike a few states away. Or nearby power lines, TVs, cell phones, certain radio waves from radio stations, hair dryers, and electrical wiring. What they look for are usually spikes in the EMF meter. Though why a ghostly presence equates with a jiggle on an EMF meter is nearly impossible to explain aside from "it happens." There are no randomized controlled trials of EMF meters and paranormal phenomena. They can't exist, because you need something real (like ghosts?) to measure. And data that is reproducible. And consistent ways to measure them.

Instead, we have photos often taken with smudges or specks of dirt on the lens, resulting in ghostly "orbs." Temperature drops can occur for a variety of reasons in an area due to drafts and air circulation. And there is no consensus on how to use an EMF meter in ghost hunting, or why temperature drops, for example, are used as evidence. And finally, it must be said that those ghost-hunting shows have never definitely proven they've found a ghost.

Banshees on the Brain

The combination of anecdotal evidence and personal experience tends to be a convenient pairing when explaining paranormal activity. But that is not to say that either can be considered hard facts. Okay, you might've seen a shadowy figure in a dark room or heard a door slam for no reason—but was it really a ghost? Countless people have odd photographs with inexplicable images or faces. But they don't really prove anything, right?

The explanation for a given ghost sighting might actually be much simpler than it may seem: It can be found in our neural wiring. Our brain often attempts to find human faces in the chaos of our visual world—for example, when the holes of an electrical socket look like a surprised face. It's called "pareidolia" (finding meaning in meaningless things), and it's why people strive to find order in chaos. From an evolutionary standpoint, humans receive a survival benefit from creating organization and finding meaning in our worlds. It's why we take more comfort in saying that inexplicable knocks and

They're everywhere.

moans in the attic are from ghosts, versus being okay with not knowing the answer. It's why we see a human face on the surface of the full moon, or a ghostly one in the dark shadows of a random photo.

Also, remember that human vision has a lot of tricks up its sleeve. We are prey to something called selective attention. By concentrating on one thing at a time, we tend to miss all the other objects or people nearby because our brain can't see everything all at once. After a busy day out of your house, can you remember what every person you saw was wearing? Probably not. But your eyes landed on those people. Our brains literally fill in details that aren't there. Right now, you have a blind spot in each of the retinas of your eyes, due to the absence of photoreceptive cells where your optic nerve passes through your optic disc, but you can't tell those spots exist, because your brain has kindly filled them for you. Our human senses are actually the worst ways to gather hard data.

One fascinating medical complication that causes people to see ghosts or other paranormal aberrations is a phenomenon called Charles Bonnet syndrome, in which a person with poor vision might hallucinate little elves or people in their living room. A floral pattern on a couch might start dancing. All because of a glitch in the visual system that has nothing to do with psychosis or munching on hallucinogenic mushrooms.

Sleep paralysis is also a frightening experience that many people have confused with seeing ghosts. It's a temporary loss of muscle control that occurs just before falling asleep or upon waking, during which time the brain is dreaming. Temporarily paralyzed, the person can't move a muscle or scream, while seeing some very unreal things. It's terrifying and could explain many so-called paranormal experiences. "The Night Hag" is a term some have attributed to a demonic being they believe holds a person down in these instances. In ancient Mesopotamia, it was called lilitu; in Thailand, it has been called phĭĭ am; and in Turkish culture, a basty. A similar and terrifying entity exists across many other cultures.

Another medical reason for feeling like you're experiencing an uncanny presence nearby could be carbon monoxide poisoning. Many stories have been reported by people who thought their homes were haunted only to find they were being gently poisoned by a CO_2 leak. CO_2 is an odorless, colorless gas produced by the incomplete combustion of fuels, and broken furnaces or old houses with a CO_2 leak make inhabitants feel groggy, confused, or dizzy, and they can experience hallucinations or suffer from blurred vision.

And finally, let's not forget bias. People who believe in the paranormal are more likely to say they experience a phenomenon when visiting so-called haunted places.

Can Science Prove Ghosts Exist?

The existence of ghosts has yet to be scientifically proven. It doesn't help that the ghost-hunting community can't agree on what a ghost is, why some exist, why some haunt us, why their physical existence (as it manifests on ghost photos or video recordings) manages to work with some Newtonian laws (photographic equipment can capture them as smudges) and defy others (they pass through solid walls). Some say that the undead are electromagnetic beings and have managed to preserve energy in that way (tipping their ghostly hats to the laws of thermodynamics).

But if ghosts exist and abide by some rules of the natural world, thus allowing themselves to be seen and detected, why can't they be proven to exist with ordinary, scientifically sound methods? And many obvious questions poke holes in the theories about the existence of ghosts. For example, why are ghosts wearing clothes? And if a ghost is, say, avenging their own murder, why don't they just tell the police? What is it about an unfortunate death that translates to the ability of a soul to exist afterward?

Then we have to consider the sheer number of undead who would be wandering among us. If just 10 percent of those who died stuck around as ghosts because they were unsatisfied with their untimely passing or unresolved issues, then Earth would be packed with the undead. They'd be literally everywhere. You'd never truly be going to the bathroom or showering alone. If that were the case, wouldn't they try a little harder to make themselves known in places that weren't dark, dreary, and deserted? Isn't it a little convenient that dark places—which happen to be when human vision is poorest and most susceptible to influence—are the usual "haunts" of ghosts? Finally, why are their voices garbled when they're recorded? Repeated photography, video capture, and audio should be able to pin them down definitively, no?

Your ghost tour awaits in places like Savannah, Georgia, reportedly rife with specters.

Yet they can't. Because they're not real.

As for the pursuit itself, ghost hunting falls squarely into the realm of pseudoscience. It resembles science in its use of all this equipment and detailed rules for the detection of ghosts, but it's missing key components of true science. First, ghost hunters are not really looking for ghosts but for aberrations on the tools they're using—and arbitrary aberrations at that. They look for them over and over again, but observation alone is not the same as empirical research, which requires validated tools of measurement. Second, there's no scientific method happening, no testing of a hypothesis in a way that is reproducible and reportable in scientific literature. Stories, hunches, interesting findings, and anecdotes might point to a trend that requires investigation. But then ghost hunters never take the next step of actually testing that hypothesis. And possibly the most problematic part of ghost hunting is that there is no solid definition of what a ghost even *is*.

Despite the lack of true data, there are a few robust drivers of the ghost-hunting movement. First, there's a lot of money to be made when it comes to haunted places. Tourism dollars convince many to never reveal the truth. In cities that are notable for their hauntings, such as Ellicott City (in Maryland), New Orleans, and Savannah, there's a robust industry built around ghosts. Then there is the equipment used by intrepid ghost hunters. You can purchase a "couples ghost hunting kit" online for an eye-popping $300, if you're part of a romantically inclined paranormal investigative team. Or perhaps the fuzzy, cute "BooBuddy" teddy bear (only $359!), which acts as a trigger for ghosts and will ask questions while picking up changes in energy, motion, and temperature.

But maybe ghost hunting isn't really about hard science dedicated to proving the existence of ghosts. Instead, it could be a kind of comfort, a way to weakly prove that our lives aren't finite and that loved ones might not be forever lost once they've died. Pointing an EMF device is akin to blurring our eyes to reality as it is—and the reality of death and loss is harsh for so many of us.

Consider Mary Todd Lincoln. We know she sought to speak with her dead son once via a séance. Despite hearing about the fraudulence of spirit photographs, she sat in Mumler's studio and was eventually handed a photograph showing Abraham Lincoln standing behind her, his hands resting gently on her shoulders. Deep in her sadness, Mary Todd could not be persuaded that her husband was truly gone.

So next time you think you see a figure in the shadows out of the corner of your eye but turn to find nothing there, know that you're not alone in wondering. Nor will the search for ghosts end anytime soon. People will keep searching gloomy, creaky old houses and enjoy the jump scares that jolt our dopamine levels into a pleasurable overdrive. Families will strive to connect with the spirits of their ancestors, with all the reverence they deserve. And in the process, we might glance into the murky past and maybe not feel so alone in this mortal realm.

Cryptozoology

n 1958, in Northern California's Humboldt County, Gerald "Jerry" Crew arrived at work like it was any other day. Crew was a cat skinner, working a bulldozer for logging companies. The area of Northern California where he worked was beautiful, with dense patches of spruce, pine, and fir trees that cut lush, green zigzags on the hills set against the August sky. Crew was a sober, churchgoing man, devoted to his hometown and his family. He worked hard, as he was prepared to do that day in 1958. When he arrived at his bulldozer, he put on his boots and hard hat. As he did, he briefly noticed that there were footprints around his bulldozer, which he didn't think much of at first until he got up into his machine and took a good look. They were odd, not like animal prints but human-like, only huge, far larger than a human's foot and sunken into the mud and earth. Whatever made them must have been enormous.

Crew called over his foreman, Wilbur "Shorty" Wallace, as others gathered around. Twenty-five other workers said they'd seen similar tracks

For your safety and a requisite Instagram post.

along nearby Mad River. Some had noticed the footprints closer to the coastline, in the city of Trinidad. The summer before in an area down the road, an enormous 450-pound drum of diesel fuel had inexplicably been moved into a gully far away, with those same enormous footprints nearby. The workers had a name for whatever was moving secretly around their worksites.

Big Foot.

A month later, more tracks appeared near Bluff Creek Road. The men who found them were used to working in the deep woods among wild animals, but these were not normal animal tracks. And they were certainly not made by a bear, nor were they fakes. Crew had noted an additional oddity: There were too many tracks to be faked, and the details too fine to be created by some kind of fabricated monster foot. A taxidermist told Crew how to make a plaster cast, and that is how he made it into the newspapers, his earnest, bespectacled face holding the plaster cast of a bare, primate-like foot that was far too large to be human.

The chances are good that you have heard of Bigfoot, as it's spelled today. Usually characterized as an uncivilized version of Chewbacca, it has lumbered in the corners of our vision and imagination for decades. You might know Bigfoot as Sasquatch, the anglicized version of *sasq'ets,* meaning "hairy man" in the language of the First Nations peoples of coastal British Columbia. It's tall by definition—anywhere from 6 to 15 feet in height. It walks on two feet like a human and is covered in shaggy fur of various shades. Some say Bigfoot's eyes glow yellow or red. Others claim that it smells like rotten garbage and dead animals. And, of course, as its name implies, it's got big feet—24 inches long, according to some.

In 1967, a pair of amateur filmmakers, Roger Patterson and Robert Gimlin, went to an area of Northern California where Bigfoot sightings had been reported. Along a creek with its crow's nest of a logjam, they saw it—

a tall, upright thing of about six feet or more with prominent breasts. The creature walked steadily away from them, glancing over its shoulder with apparent irritation before eventually disappearing into the woods. Having written about Bigfoot in the past, Patterson just so happened to also be filming a docudrama about the creature when he shot the historic film that day. He and Gimlin received plenty of attention and money from the film. Although the footage was mostly chaotic and unsteady, many people still believe it to be definitive proof that so-called Bigfoot exists. Both Patterson and Gimlin insist they didn't fake the footage.

The iconic Bigfoot filmed in 1967, annoyed by paparazzi

But many others dismissed it. Some of those who did theorized that it was a relative of the *Gigantopithecus*, a massive primate that once crossed from Asia to North America. This type of theorizing opened up a field called "cryptozoology," a term that took off in the late 1970s. It is the study of "cryptids," or animals that aren't recognized by scientists as real but are believed to exist somewhere. Even if *Gigantopithecus* died out a million years ago, cryptozoologists wondered if it may have survived in pockets of wilderness somehow. And *if* Bigfoot were a type of ape, rather than a human in a hairy costume, it would have certain features similar to other apes. Compared to modern gorillas, for example, this Bigfoot had breasts and buttocks that were too hairy. Also, the front-to-back bone on the top of its head, known as the sagittal crest, was too prominent. In short, this Bigfoot didn't make sense.

If you're of a certain age—Gen X and older—you might remember the brief bit of Patterson's grainy Bigfoot film that opened up the television show *In Search of*... narrated by Leonard Nimoy's authoritative, überlogical Spock voice. It was entertainment, sure, but millions of viewers were glued to the screen in search of the truth. The fantastical. The mythological that was actually real.

Beyond Bigfoot: Champy, Nessie, Mothman, and More

Every country has its own cryptids, some fairly famous. There's Nessie, the Loch Ness monster. Another is the chupacabra, a fanged creature with glowing red eyes that sucks goats dry of their blood and is supposedly native to Puerto Rico and South and Central America. Then there's the cold-climate version of Bigfoot, the Yeti or Abominable Snowman, which roams the Himalayas.

The Loch Ness monster photograph from April 19, 1934, was said to be taken by "a surgeon," but Chris Spurling, one of its creators, later admitted on his deathbed that it was a hoax.

There are others, too, ones that are not as well-known. For example, the Ningen. In the subantarctic region, rumors say it has a pale blue humanoid face with no torso, just legs, or something more whale-like with flippers or fins. Or perhaps you've heard of the Mongolian death worm? It's a creature that lives in the Gobi Desert and is said to be a two-foot-long, sausage-like worm that kills humans on contact.

Heck, just about every state in the US has a cryptid all its own. For starters, Bigfoot isn't just a California thing—it's been sighted in every state. State-specific cryptids can be fairly famous, like Champ, the monster of Lake Champlain, which borders Vermont and New York, and is said to resemble a large aquatic dinosaur, like Nessie. The Loveland Frogman in Ohio appears in illustrations as a man-size frog that walks on its hind legs. Some states love their cryptids so much they name their hockey teams after them (New Jersey, for its Jersey Devil). Don't overlook the oft-forgotten Goatman of Maryland, the Oklahoma Octopus, the Virginian Bunnyman, or the Honey Island Swamp Monster of Louisiana.

But if there's one place to get the most cryptid bang for your buck, it's West Virginia. There, the state lovingly lays claim to the Flatwoods Monster, Bigfoot, the Mothman, the Grafton Monster, Sheepsquatch, the Snarly Yow, the Cumberland Dragon, the Abbagoochie, Preacher Bat Boy, and the Vegetable Man. In fact, the official West Virginia government

CRYPTOZOOLOGY | **129**

TOP LEFT: *The Mongolian Death Worm: To merely touch it means instant death, though no one has ever seen one.* TOP RIGHT: *The Loveland Frogman. Part man, part frog, all disturbing.* BOTTOM: *The Mothman, haunting your West Virginian dreams and making bank at the annual Mothman Festival.*

website even has a page dedicated to them, and the state hosts a Mothman festival every year.

For every cryptid, there exists many more cryptozoologists who have spent countless hours researching and tracking down these creatures, looking for evidence of their existence. And for every cryptozoologist, there is a zoologist who says that cryptozoology is pseudoscience, plain and simple. After all, it does carry all the hallmarks of pseudoscience—including claims that are unfalsifiable, a failure to adhere to the scientific method, a distrust of the mainstream scientific community, and the questioning of repeated evidence that disproves its beliefs.

Cryptozoology is a more contemporary manifestation of what humans have been doing for a long time: attempting to catalog and verify the existence of new and recently discovered animals, as well as rumored ones. The *Liber monstrorum* from the seventh or eighth century included pygmy peoples and hippopotamuses, as well as minotaurs and dragons. In the medieval world, Albertus Magnus wrote *De animalibus*, a thirteenth-century compendium of animals that included their habitats, behaviors, and physical descriptions. It was far more scientific than the bestiaries of the day, which often included fantastical creatures like unicorns and mer-peoples and had a tendency to describe them within a moral context. For example, the beaver was sometimes depicted biting off its own testicles, which were thought to be connected to the castoreum glands, a valuable commodity at the time, to save itself from a hunter. This was not only awful—it was impossible, because the beaver's testicles are internal. Perhaps the visual alone functioned as a lesson to medieval readers about survival. Often, the more fantastical bestiaries weren't trying to perpetuate untruths—they simply didn't possess the ability to prove or disprove the existence of these creatures. Interestingly, the prologue of *Liber monstrorum* warns readers, "*sequentem historiam sibi quisque discernat*," meaning that they must evaluate each story for themselves in terms of what to believe or not.

Medieval illustration of a beaver (yes, that's supposed to be a beaver) displaying impressive flexibility while biting off its own testicles

After the Age of Exploration, information about animals from formerly geographically isolated parts of the world was disseminated widely through illustrations and written accounts. Medieval bestiaries depicting creatures evolved into compendiums of zoology, botany, entomology, and mycology specimens by European naturalists like Gilbert White and William Kirby. Moving forward and well into the nineteenth century, there were fewer and fewer lands to explore, and the animals of the most remote regions of the Earth were soon revealed. Legends of sea serpents taking down ships seemed less and less believable as the years went on. Our once extraordinary world became ordinary. The hidden recesses were all but revealed. And then we began to systematically destroy these animals and their natural habitats. The passenger pigeon. The dodo. The Tasmanian tiger. The Steller's sea cow. Humans caused a mass extinction rivaling that of the one that took place 66 million years ago (and without a massive asteroid). Much of the world remained ignorant to the damage that would come from the destruction of habitats and culling of wildlife on massive scales.

The idea that there are animals yet to be discovered is a delicious fantasy. What is dearly missed is a time when the destruction of habitats and culling of wildlife wasn't yet such an obvious horror show. Perhaps we wish for that moment when we didn't have all the answers, and when new discoveries were the size of whales, not nanoparticles.

Hidden . . . or Nonexistent?

Cryptozoology wasn't an actual field of study until very recently. In 1955, Bernard Heuvelmans, a Belgian zoologist, published a book called *On the Track of Unknown Animals*, while a Scottish zoologist, Ivan T. Sanderson, published *Abominable Snowmen: Legend Come to Life*. The term "cryptozoology" (Greek for "the study of hidden animals") was apparently coined by Sanderson. But modern zoologists, who you might think would be interested in discovering new species of insects or animals, shied away from the term and those who identified themselves as cryptozoologists.

Cryptozoologists were not systematically pursuing cryptids in a way that scientists might—which is to say, with consensus, reproducibility, and validation. Or even, these days, with DNA evidence. They tend to use things that would sound familiar to ghost hunters (see page 112), such as audio recording equipment, night vision goggles, and photography. They also exhibit the familiar tendency to shun mainstream scientific methods, including fossil records and evolution. But the problem with calling out cryptozoology isn't as simple as saying their methods don't work. As long as you can't actually find the cryptid—and, for example, produce a carcass to study that is agreed upon by zoologists as real—you can't say it doesn't exist either. Cryptids can still live forever in these double negatives of possibility, in the ether of the imagination, and, of course, in those mist-laden, creepy woods of rural West Virginia.

To be fair, there are several examples of mythical monsters made real for European eyes once they were legitimized into existence. Gorillas, well-known to the Indigenous peoples of East Africa, were likely first seen by English sailor Andrew Battel in 1590. He wrote that the *"Pongo* [*Mpungu*] . . . is more like a giant in stature than a man for he is very tall, and hath a man's face, hollow-eyed, with long hair upon his brows . . . [they] are never taken alive, because they are so strong that ten men cannot hold one of them." It wasn't until centuries later, in the mid-1800s, when gorilla remains were brought to England, that these creatures were taken out of the realm of the faraway and fantastical for Europeans. It was probably hard to believe kangaroos were real, too, with their fox-like heads, monkey tails, belly pouch for their young, human-like hands, and the powerful legs of a ginormous bunny. Then there's the platypus. In 1798, when the animal was first seen by European colonists on the east coast of Australia, pelts and sketches were sent back to scientists in Great Britain, who were still skeptical of its veracity. Even biologist George Shaw, when given a preserved animal, thought it a clever fake. He believed the beak was stitched on by a taxidermist and took it apart with scissors to look for the telltale stitches. But it soon came to be that the animal seemingly cobbled together by a toddler's imaginings—a duck-like beak, brown fur, a beaver tail, egg-laying capacity, and venomous claws—was indeed a real animal, one that left the gray shadows of the cryptid world for reality.

The giant squid is another cryptid that was once most definitely a thing of nightmares, hearkening back to medieval and nineteenth-century drawings of giant octopuses by European and Japanese artists. There were tales

This giant octopus would like a word about your sushi order.

of fearsome sea creatures like the Norse kraken, which would sink ships with nightmarish violence. Pliny the Elder and Aristotle mentioned the giant squid in their writings. Sightings of beached and dead squid arose throughout the last several centuries, but no specimens survived preservation, making it harder for those who hadn't set eyes upon the beasts to believe they were truly real. For something so colossal—upward of 43 feet long, according to some lore—it was hard to accept them as anything but legend. It wasn't until the late 1800s when photographs of dead giant squid, along with more documentation of beachings, made it clear that the giant squid was not a nightmare after all. In fact, it was only recently, in 2002, that a live giant squid, *Architeuthis dux*, was photographed swimming off the coast of Japan. The first video of an adult giant squid in its deep-sea habitat was filmed in 2012.

To be clear, an undiscovered species is not the same thing as a cryptid. Nevertheless, many new species are identified constantly now, ones whose existences continue to surprise us. Entomologists, for example, come across never-before-documented species every year, like the *Neuroterus valhalla*, a wasp discovered outside a Rice University student pub, or the caddisfly, found in Kosovo. In the deepest ocean trenches, marine biologists are still discovering new sea cucumbers, feathery stars, bulbous-eyed fish, and never-before-seen poofy octopuses. Let us also not forget the recently discovered COVID-19 virus, or the more than 10,000 previously unknown viral species that were discovered living in the poopy diapers of Copenhagen babies in 2023. (Eww, but also . . . cool?)

The famous Feejee mermaid, possibly suffering from a killer migraine

Generally, cryptids have storybook qualities to them. They are creatures who have been reportedly seen but never with definitive proof. They are often larger than, say, dogs. And they tend to have some very striking, strange, and sensational traits that make them particularly fearsome. There are also typically arguments over whether or not they really exist. They are usually ignored by mainstream zoologists. And, finally, they stubbornly refuse to be found.

But this fact has not stopped people from trying. For the Loch Ness monster, sonar surveillance has found only shoals of fish or inconclusive blobs. Photos are often easily explained by driftwood, waves and wakes, waterbirds, swimming deer, buoys, boats, or fuzzy images of nothing at all. Diving expeditions have also failed to find the creature. At least 10 or more Nessies would need to exist to maintain their ongoing population. This is according to a 1972 paper debating the population density using calculations that include estimated weight. ("The minimum average size is taken arbitrarily as 100 kg; anything smaller is not suitably monstrous.") Another fact is that no carcasses have ever been found. An environmental DNA sampling of the Loch Ness waters revealed DNA, not of a modern-day plesiosaur, but of a lot of eels. Which may be what Nessie actually is—an eel, or perhaps a diving swan that got caught by a lucky paparazzi camera angle, as might have been the case in a famous 1933 photo. Or possibly a toy submarine topped with a plastic replica of a creature's head and neck, which was the "monster" in the most famous photograph of Nessie. The picture was revealed to be a fake in 1994, 60 years after it was originally created.

In the cases where evidence is presented, it's often found to be a hoax, like the Feejee mermaid, a dried-out specimen consisting of the tail of a fish sewn onto a shrunken monkey torso that P. T. Barnum displayed in his 1842 American Museum. It later disappeared, likely a victim of the many fires that hit his collections. Sir Edmund Hillary, famous for being the first to summit Mt. Everest along with Tenzing Norgay, went on an expedition in 1960 to discover the Yeti, or Abominable Snowman. They brought plenty of rifles, tear gas, and shotguns, hoping to capture one alive. They did find enormous footprints, but realized they were likely human footprints enlarged by the action of melting snow. Disappointingly, the uncovered so-called Yeti fur turned out to belong to the Tibetan blue bear.

Fake Creatures, Real Places

It's fascinating that the giant squid can grow to such monstrous proportions. The reason it can is something called *abysmal gigantism*, where creatures that live in deep water and under high pressures tend to grow to larger proportions than their more shallow-living cousins. But it's also in those environments that are so wholly hostile to the existence of humans that they can flourish. Which is one major hallmark of the cryptid. It won't be living in downtown New York City but in the rural wild reaches of the world. And that's a romantic notion, isn't it? Movies know how to tap into our desire to discover the fantastical in our ever more ordinary world, as happens in *Splash*, when a cryptozoologist is finally legitimized upon the capture of Daryl-Hannah-as-mermaid. Or in any one of the *King Kong* movies, with that island full of monsters that filled the movie screen—and our nightmares.

We must also remember that it's the cryptids that have placed many obscure towns on the map. Tourism brings in money to places that would otherwise be forgotten. A small town like Whitehall, New York, proudly displays its many Bigfoot statues. Sutton, West Virginia, has a quaint Flatwoods Monster Museum, where you can buy a Flatwood Monster ceramic lamp to bring home and light the darkness with its weird robotic head. Or you could go to Fouke, Arkansas, for the Fouke Monster Festival that celebrates its own shaggy, Sasquatch-like beast every year.

* * * * *

Cryptozoology comes with an enthralling, bittersweet relationship with the beings it seeks to prove into existence. As Friedrich Nietzsche wrote, "The real world is much smaller than the imaginary." Cryptozoology is an enticing concept in this overexplored, overexploited, and forever Instagrammed world, one that can perhaps still hide pockets of creatures that defy being caught, alive or dead.

But it's not just the allure of imaginary realms that draws people to cryptids. It's also about being immersed in the last few mysteries that still exist in our world—and true mysteries are more at risk than ever in our science-based yet consumeristic lives. If we're destroying habitats, we might also be destroying cryptids systematically, by encroaching on their territory. There have been calls to pull the cryptozoologists' fervor into environmental conservation, as maintaining biodiversity is in the best interest of cryptozoologists, biologists, and ecologists, too. There remains a hope that our long-running fascination with the wild and improbable can do some good before we kill not just more known species around the world, but maybe also those that inhabit our dreams.

The mystique endures in the meantime. Maybe it's better to think of cryptids and cryptozoology less in the bubble of pseudoscience and more as a cultural phenomenon that speaks to our need for discovery and jump scares in the haze of scary stories. The love of cryptids also remains relatively harmless, infuses income into tiny towns across the country, and brings people together through the folklore of their local communities. Thankfully, we don't need to traverse the leech-filled waters of the remote Amazon to find these hidden creatures. After all, Bigfoot has been spotted in every US state—you don't have to look a heck of a lot farther than your backyard to begin the hunt.

2012 Phenomenon

Booking a room for your vacation is one thing. Reserving a room for the apocalypse is something else altogether—something that takes advance planning. A *lot* of advance planning.

By early 2011, Sigrid Benard, an innkeeper in Bugarach, France, was overwhelmed with phone calls from tourists hoping to book one of her rooms from the beginning of December 2012 until the end of January 2013. But these callers weren't just tourists getting an early start on planning their European vacations. Rather, the unhappy locals called them "esoterics," or people from all over the world hoping to ride out the 2012 "Maya apocalypse." In an interview with the *New York Times*, Benard said, "People know I'm closed in the winter . . . but those people said they wanted to come three weeks before the apocalypse and book the week afterward to see what happens."

Bugarach is a small village of about 200 people in the Aude district of southern France at the foot of an unusual mountain bearing the same name. The mountain is a bit of a geological oddity. It's known as the "upside-down mountain," because the rocks found on lower elevations of the mountain are younger than those found toward the top. It's also riddled with caves and is notorious for strange sounds and odd lights. For years, it's been a draw for

Bugarach, the "upside-down" mountain

New Age believers who feel the energy is "different" there. UFO enthusiasts even believe aliens live underneath the mountain.

For some, it also seemed like the ideal place to weather an apocalypse. According to various internet theories, the healing energies of the "sacred" mountain could protect you from the end of the world. Barring that, those aliens living underneath the mountain would benevolently whisk away any nearby survivors. A third, less-thrilling possibility was that the "end of the world" would just be a symbolic transition to a new, better age. If so, what better place to watch the world blossom under a sudden flowering of peace and love than a sacred mountain in southern France?

Back to Benard's problems with booking her rooms. Starting with: What exactly was the apocalypse everyone was trying to escape ... or to witness?

The winter solstice, December 21, 2012, marked the end of the Maya Long Count calendar. But what did an ancient Maya calendar have to do with doomsday preppers and UFO believers? To many people, the end of the Long Count calendar meant that the world was doomed and so the apocalypse was certain. Or, at a minimum, human consciousness was primed to enter a new age. Regardless, they felt that one thing was certain: *Something* was going to happen.

December 21, 2012, was not a date to overlook.

The Long Count to the Apocalypse?

The range of eschatological beliefs that surrounded December 21, 2012, as the date of the "Mayan apocalypse" stemmed from misunderstandings of, and misattributions to, the Mesoamerican or Maya Long Count calendar.

In it, December 21, 2012, marked the conclusion of a cycle of 13 baktuns, a Maya astronomical measurement of time equal to approximately 394 years. According to ancient Maya mythological traditions, the world in 2012 was heading toward the end of the "fourth age." On December 21, 2012, the Long Count calendar would reset to "zero" at the conclusion of the 13th baktun, thus starting a new "fifth age."

The Mesoamerican Long Count calendar is a linear construction of time, likely invented by the Olmecs, the first major civilization in Mexico, before being embraced by, and then permanently associated with, the Maya civilization as early as 300 BCE. The Long Count is an astronomical calendar, determining universal cycles that, according to Maya mythology, may have marked the beginnings and endings of world ages. It's a little unclear what exactly the ancient Maya thought happened at the beginning or ending of Long Count calendar cycles. The Long Count age they inhabited, which was the same "fourth age" that people inhabited in 2012, had begun on August 11, 3114 BCE, a date traditionally associated in Mayan mythological beliefs as the day human beings were created.

The Long Count calendar in turn interacts with two other calendars used to identify days, the Tzolk'in (or "divine") calendar and the Haab (or "civil") calendar.

The Tzolk'in calendar is a cyclical 260-day calendar, with 20 cycles of 13 days called "trecenas" that constitute one year in the Gregorian calendar. This calendar was used by the ancient Maya to identify and mark important religious and ceremonial events. The Haab calendar, meanwhile, is a 365-day cyclical solar calendar with cycles comprising 18 months of 20 days (known as an "uinal") and 1 month of 5 days

The Maya calendar

(known as an "uayeb"). When combined, the Tzolk'in and Haab calendars form what is collectively called the "Calendar Round," which repeats in 52-year cycles. This construction of time is still used in parts of Guatemala with deep Maya heritage, particularly in the Guatemalan highlands.

The Long Count, or astronomical calendar, was added to the Maya conception of time sometime around 300 BCE. It introduced the concept of baktuns to the calendar.

The Tzolk'in, Haab, and Long Count calendars all interlocked together in a rather beautiful construction and understanding of the passing of time. Twenty days made a uinal, 18 uinals (360 days) made a tun, 20 tuns made a katun, and 20 katuns (144,000 days, or approximately 394 years) made a baktun.

Unlike the cyclical Tzolk'in and Haab calendars, however, the Long Count calendar is linear. Thirteen cycles of baktuns (roughly 5,125 years) marked the end of a "world age," when the calendar was reset to a zero.

And here's where it gets apocalyptic.

In the *Popol Vuh*, a sacred text of the Maya people, the K'iche' Maya from the Guatemalan highlands recorded their creation myth. The surviving text describes the Maya gods creating three failed worlds before successfully making a fourth world where humanity was placed. The last "world age" had ended after a cycle of 13 baktuns. By the same reckoning of 13 baktuns per world age, the fourth world would come to an end at the conclusion of its 13th baktun . . . on December 21, 2012.

Predictably, once the news spread on the internet that December 21, 2012, would be the end of a "world age" according to ancient Maya cosmology, a whole mess of other gibberish and misinterpretation got mixed in.

Fast.

The Beat of the Baktun

Soon, there were notions cropping up on the internet that Earth would collide with another planet or another large cosmic object on December 21, 2012, or that magnetic poles could reverse, or a black hole would appear over the Earth and swallow it whole.

But it wasn't just pseudoscientists who debated the potential for an apocalypse. Mayanists (scholars who specialize in studying the ancient Maya) also wrote about the significance of the end of a baktun. The first reference appeared as a footnote in astronomer Maud Worcester Makemson's 1951 translation of a book of Maya prophecies: "The completion of a great cycle of thirteen baktuns . . . would indeed be an occasion of the highest expectation." (As apocalyptic predictions go, it's admittedly underwhelming.)

Then in 1966, another Mayanist, Michael D. Coe, wrote in his book *The Maya* that "there is a suggestion . . . that Armageddon would overtake the degenerate peoples of the world and all creation on the final day of the thirteenth baktun." (Now we're getting somewhere!)

While later Mayanists would refute these apocalyptic claims, popular American writers including Frank Waters and Terence McKenna began to actively promote them. Soon these claims sank into the public

Just a flesh wound

consciousness, roped into a general circle of New Age beliefs about the ancient Maya people that is collectively referred to as "Mayanism." With roots stretching all the way back to early New World exploration, when many Europeans were swept up in romantic ideas of "lost civilizations," Mayanism was revived in the 1970s largely through the popularity of Frank Waters's writing and those who followed in his wake. This New Age Mayanism, lacking any central belief, tends to imbue the ancient Maya with understandings of consciousness and cosmology far in advance of modern science.

In 1975, Frank Waters published a book entitled *Mexico Mystique: The Coming Sixth World of Consciousness,* which really got the apocalyptic ball rolling. After incorrectly overestimating the duration of a 13-baktun cycle (5,200 vs. 5,125 years), he wrote about five legendary eras corresponding with five baktun cycles, each of which ended with the destruction of Earth and its subsequent rebirth. (None of this was based on actual Mayan sources; rather, it was a heap of speculation piled onto Waters's limited grasp on Maya conceptions of time.) Waters wrote about the distinct possibility of Earth being destroyed by "catastrophic earthquakes" on December 24, 2011. (Not a typo—he also got the date wrong.) But after balancing his interpretation of Maya time conceptions with a heap of Western astrology, Waters softened his claims for 2011, writing instead "the general prognosis is good. The world will not be destroyed by a great cataclysm despite all dire prophecies. Nor will mankind. Both will continue their evolution in a new phase. A phase significantly different than the one we are now experiencing."

The Age of Aquarius.

Other authors picked up on these threads, including Terence McKenna in his book *The Invisible Landscape* and José Argüelles in *The Mayan Factor.* Soon, the apocalypse concept had leaked into public consciousness. Before you could say "ancient Mayan

Terence McKenna

apocalypse," 2009 rolled around and we all flocked to theaters to watch a Roland Emmerich–directed disaster flick entitled simply *2012*. While watching John Cusack run away from an exploding volcano in Yellowstone National Park has a certain appeal, the movie mostly focused on the pseudoscientific theory of crust displacement (see sidebar, page 146) and paid only minor tribute to actual Mayan sources.

Zetans to the Rescue!

Some New Age writers followed in Waters's footsteps and thought the coming apocalypse was spiritual. The conclusion of the 13th baktun meant that human consciousness was about to enter a new age. Others, well, went in a different direction. Soon, internet sites began to spread fears that the planet Nibiru would collide with Earth. Never heard of Nibiru? That's because it doesn't exist. Nibiru, or "Planet X," was believed by some at the time (before Pluto's demotion) to be the 10th planet in the solar system.

In 1976, Zecharia Sitchin, a shipping executive brimming with confidence after having taught himself to read Sumerian cuneiform, wrote a book about a shocking discovery: There was another planet, Nibiru, that orbited the sun every 3,600 years. Apparently, the ancient Sumerians, lacking any sort of astronomical technology, knew about a planet that had escaped the attention of NASA.

While Sitchin was content to leave it at that, Nancy Lieder, founder of the website ZetaTalk, originated the Planet Nibiru collision course theory in 1995. She claimed on her site that she was contacted by a group of aliens called Zetans, who had, rather thoughtfully, implanted a special communications device in her brain so she could understand them. According to Lieder, the Zetans were an advanced race of extraterrestrials from the Zeta Reticuli star system.

In a continuing effort to be helpful, the Zetans informed Lieder through the device in her brain that a large planetary-sized object (Nibiru!) would sweep through our solar system, igniting a polar shift that would destroy life on Earth as we know it. The idea was that a passing planetary body, even if it didn't collide with Earth, would wreak havoc on our gravitational field and initiate a pole reversal.

Here come the Zetans.

Boldly, Lieder initially declared the apocalyptic date as May 27, 2003, but when May 27, 2003, passed without incident, Lieder wised up and realized that all the best apocalyptic leaders never make the mistake of committing to a certain date.

After declaring May 27, 2003, was a date designed "to fool the establishment," Lieder refused to disclose the actual date of the pending apocalypse. (Naturally, one might wonder how the helpful Zetans must've felt about her withholding this information.)

Not to worry, though, because other people on the internet were more than happy to provide a new date for the apocalypse in the wake of Lieder's inability to commit. And the date the internet chatter landed on, of course, was the end of the Maya baktun on December 21, 2012.

Polar Shifts and Solar Maximums

While the thought of a 10th planet abruptly appearing in the night sky on a collision course with Earth seems absurd to most people, the potential impact of a large celestial body passing by Earth is worth considering against the science. The planet Nibiru apocalyptic theory primarily relied upon fears of a reversal of Earth's magnetic poles. But while this sounds rather unsettling, is it really something to be worried about?

Every few hundred thousand years, the magnetic field over the Earth does indeed reverse. (This means that, 600,000 years ago, your modern compass would tell you that you are facing south when you are facing what we today would now call north.) In the course of Earth's three-billion-year history, the poles seem to have reversed on average once every million years or so. But it's not like you wake up one day and north is south. The reversal happens gradually, by human lifespan standards, over the course of hundreds or thousands of years. During that time, magnetic fields "morph and push and pull at one another, with multiple poles emerging at odd latitudes throughout the process," according to NASA. And while that does sound odd, would it be actually life-threatening? The last time the poles reversed, *Homo sapiens* weren't yet capable of recording their experience, but scientists who've studied the geological and fossil records of hundreds of past magnetic polarity reversals have resoundingly said we have nothing to worry about.

A related fear in 2012 was that a pole reversal would leave Earth without a magnetic field while the poles were busy reversing themselves. The Earth's magnetic field protects us from solar flares and coronal mass ejections (large explosions of plasma) from the sun. People were concerned that a solar maximum (when the sun is at its most active in an 11-year solar cycle, with more frequent and intense sunspots and solar flares) would be particularly deadly if Earth was temporarily without a magnetic field. As is so often the case with such theories, some real science got mixed up with some gibberish to stoke the flames of fear.

It's true that solar activity has a predictable cycle that peaks every 11 years or so. It was also true that one such peak, or solar maximum, was

> ## Earth's Crust Displacement
>
> Is Antarctica actually... Atlantis? After World War II, Professor Charles Hapgood took a position at Springfield College in Massachusetts, where he would teach history, anthropology, and the history of science for several decades. One day, his otherwise promising academic career took an abrupt turn toward the pseudoscientific. It was all because a student asked Hapgood, "Whatever happened to the lost continent of Mu?" (Nothing happened to Mu, because it never existed; however the idea of this missing continent in the Pacific Ocean was popularized in the early-twentieth-century works of James Churchward.) Hapgood took the question seriously and painstakingly developed "crust displacement theory" to explain how a continent could, potentially, disappear.
>
> Basically, the theory laid out supposed evidence of dramatic, catastrophic (apocalyptic?) shifting of the Earth's crust at various times in relation to its axis. According to Hapgood, this phenomenon occurred several times over the course of Earth's history, and when it did, whole continents were rearranged... or lost entirely.
>
> It's a theory long since discredited by the study of plate tectonics and continental drift. But for a brief period in the mid-twentieth century, it was taken seriously. In fact, Hapgood was able to get none other than Albert Einstein himself to write the introduction to his book *The Earth's Shifting Crust*, which was published in 1958.
>
> It turns out, even geniuses can't be right all the time.

predicted for 2012 to 2014, roughly coinciding with the end of the Maya baktun. However, the impact on Earth was never anticipated to be more than minimal. And by minimal, NASA literally meant just minor interruption of satellite communications—a far cry from dogs and cats living together and blood running in the streets. The solar maximum was just another predictable cyclical peak in solar activity.

For the sake of argument, what if this solar maximum *did* happen to coincide with a pole reversal? Well, as we have already learned, a pole reversal is hardly an overnight process. And there's no evidence that any of the previous pole reversals (and remember, we've been through many) have ever left the planet without a magnetic field.

Although it's true that the Earth's magnetic field can weaken and strengthen over time and a weaker magnetic field would lead to a (small) increase in solar radiation on Earth, our thick atmosphere does an ample job of protecting us against the sun's incoming particles. According to NASA, the primary impact of such an occurrence would be "a beautiful display of aurora at lower latitudes."

So you wouldn't have to pack yourself off to Iceland in the dead of winter to catch the Aurora Borealis. Win-win.

Mayanism vs. the Real Maya

The elephant in the room while the internet pulsed with 2012 apocalypse theories was the Maya people themselves, and what they thought of the whole affair. Despite what some ignorant doomsday theorists may have said in 2012, the Maya aren't an extinct civilization—more than eight million Maya live in Mesoamerica and countries throughout the world.

"We have to be clear about this. There is no prophecy for 2012," said Erik Velásquez García, an etchings specialist at the National Autonomous University of Mexico (UNAM) in an interview with Reuters. "It's a marketing fallacy."

The National Institute of Anthropological History in Mexico issued a similar statement: "The West's messianic thinking has distorted the worldview of ancient civilizations like the Mayans." The statement continued by pointing out that of the more than 15,000 glyphic texts discovered in ancient Mayan ruins, an underwhelming 2 glyphs actually mention the date 2012. And plenty of dates *after* 2012 are also mentioned. Not quite the end-of-the-world apocalypse you might have been looking for.

Indeed, the ancient Maya themselves seem to have given only some minor significance to the date. Most classic Mayan inscriptions do not make any prophetic proclamations about the end of the baktun.

Martin Sacalxot, the Indigenous peoples' advocate of the office of Guatemala's human rights ombudsman, remarked that the end of the calendar had nothing to do with the end of the world. And village elders interviewed

Public Maya ceremony conducted near pyramid

by a blogger in Belize said, "Our ancestors didn't believe in death or the end. They celebrated events like these by building new monuments, venerating ancestors, and doing festivals."

And that's exactly what they did when December 21, 2012, finally rolled around. While the New Age theorists were holed up in Bugarach and other locations, including Sirince, Turkey, that were said to have enough "positive energy" to weather the apocalypse, the Maya people were organizing festivities to celebrate the end of the baktun.

In territories all across Mesoamerican countries that were formerly part of the Maya civilization—Mexico, Guatemala, Honduras, and El Salvador—celebrations were held at Maya sites to commemorate the event. The Maya, as well as tourists, journalists, and other interested parties, arrived at places like Chichén Itzá, Tulum, and Palenque, where fire ceremonies led by traditional priests were held at dawn. During one ceremony in Guatemala at Tikal National Park, the priests asked for "unity, peace, and the end of discrimination and racism."

Now that's a new world for us all to believe in.

Wishful Thinking

Cryonics

fter Hall of Famer, 17-time All-Star, and all-around baseball legend Ted Williams died in 2002, there was far more discussion about his frozen corpse than his accomplishments on the diamond. But the first order of business was his family's legal battle over whether or not Williams truly wanted to be put in "biostasis," or preserved at –320°F in an everlasting liquid nitrogen bath.

Then of course there was the tuna can stuck to his frozen, decapitated head.

Allegedly, Williams's famous head had been stored in the Alcor Life Extension Foundation in Arizona, and the tuna was supposedly used so the supportless head could rest on *something*, thus preventing it from sticking to the bottom of the storage container. (You know, just the usual decapitated head problems.) When an Alcor employee was moving the head from one container to another, apparently the tuna can just kind of . . . stuck. Another Alcor employee, Larry Johnson, detailed the event in his book *Frozen: My Journey into the World of Cryonics, Deception, and Death*. He wrote that an Alcor employee tried to remove the tuna can by striking at it hard with a monkey wrench.

But he missed.

The wrench allegedly hit Ted Williams's frozen head "dead center" and "tiny pieces of frozen head sprayed around the room." It took one more thwack with the wrench to finally knock off the tuna can.

The lengths that people will go through to live beyond their fated years are long. Very, very long.

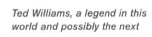

Ted Williams, a legend in this world and possibly the next

We don't live forever. Period. In the face of such a statement, you might nod and think, "C'est la vie." Or you might be filled with deep FOMO (fear of missing out) and wish you could see the future in a thousand years. Or maybe you're just terrified that your time is very limited.

Regardless of how you feel about mortality, the fear of death and the dream of immortality are both profoundly potent things. Life can feel like a seesaw. We spend our childhoods on a trajectory of growth toward adulthood. Then in a snap we slide off the apogee of youth and watch as our hair grays, our joints stiffen, and illness takes bites out of our longevity nibble by carnivorous nibble. Or perhaps your life is cut painfully short by an illness or a malignancy. Or maybe it's the possibility of the void of nothingness that follows death. There are truly countless reasons why the concept of immortality, or a second chance at life, is utterly irresistible to some. If only we could freeze time.

Freeze time, that is, by freezing ourselves.

Consider the first person ever cryopreserved: James Bedford. A psychology professor at the University of California, Bedford died in 1976 at the age of 73 from metastatic kidney cancer. Shortly after succumbing to cardiac arrest, his body was injected with dimethyl sulfoxide and Ringer's solution and then frozen in liquid nitrogen. After being moved about several times, his body currently rests in a facility in Arizona in a Dewar, or vacuum flask, not unlike what you use to keep your coffee hot for hours while commuting. Only this Dewar is quite large and holds bodies (and decapitated heads, and occasionally various pets) at −196°C, or −320.8°F. The hope? That someday, science will catch up to his dreams of being resurrected.

Mysterious cartridges containing mysterious cryogenic freezing substances

Bedford is not alone in that cold Dewar. As of 2023, about 500 people have been frozen cryogenically and about 5,000 are actively contemplating the same thing. Those who are in "biostasis" awaiting to be reanimated include doctors (L. Stephen Coles) and software engineers (Hal Finney). Many are on a list to be cryopreserved, like PayPal founders Luke Nosek and Peter Thiel. Some get cold feet, like Larry King and Timothy Leary, who both once wished to be frozen but changed their minds. The abominable Jeffrey Epstein disturbingly considered having his head and penis frozen so he could "seed the human race with his DNA."

Walt Disney and Mickey Mouse, frozen in time. In the photo, at least.

Walt Disney, founder of the company that bears his name, is rumored to be cryogenically frozen. Apparently, a reporter snuck into Saint Joseph

Medical Center in Burbank, California, and claimed to see the dead media mogul in a cryogenic cylinder. Some employees spoke about Disney's "big freeze" as if it were the truth but were later discredited. It didn't help the speculation that in Disney's last moments on camera before he died, he spoke ominously to his colleagues, telling them that he hoped to see them soon. But his family brushed away the rumors. Diane, his daughter, said in 1972 that "There is absolutely no truth that my father, Walt Disney, wished to be frozen. I doubt that my father had ever heard of cryonics." Apparently, this also dashed the rumors that his frozen body is buried beneath the Pirates of the Caribbean ride at Disneyland. The skeletons on that ride (or beneath it) are not real, we're afraid.

Human Freezing: A Most Delicate Art

In terms of cryonic details, there are many, starting with the terms "resurrected," "reanimated," and "revived." Though resurrection and revival generally mean bringing back a whole person from the dead with personality and memories intact, while reanimated technically means making the body alive again after death with a mind not necessarily intact (think: zombie movies), they are often used interchangeably. However, the concept of life after death via cryonics nevertheless means different things to different people.

Some people wish for their actual body and mind to be preserved as it existed prior to death, to experience a heretofore unknown scientific intervention to bring life back to them—all of them—mind, memories, and body fully intact. The real hope with speculative medical breakthroughs is that they would potentially reverse aging or cure previously incurable ailments that may have caused the death in the first place. Others assume that if their body wasn't able to survive this life for whatever reason (age or a malignancy, for example), then maybe the future will furnish a new body. Biological or artificial, who knows. And then all that's needed is to transfer the contents of the brain to the new vessel, via "neuropreservation"— including, ideally, with one's sense of self still firmly encased in one's noggin.

Still others believe it's not so much their frozen brain that survives but the mapping of all their neural connections, which is called the connectome.

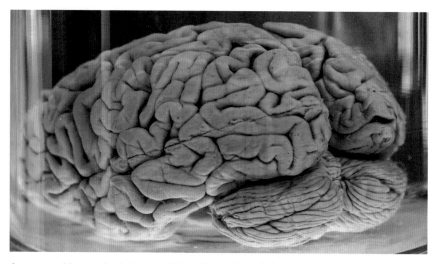

A preserved human brain. Personality no longer included.

Some think that these connections can be re-created "in silico," as it's called, or via a computer simulation. The hope is that someday, the cell-to-cell brain structure can be examined, "downloaded" (in our rough tech speak of today), and then uploaded into a new vessel. And that this would happen, once again, with a sense of self and memories intact. Sort of like the human body equivalent of getting a new cellphone and transferring all the contacts and data from your old one into it.

As you might imagine, there are several flaws with all these concepts. The cells that make up a brain and your nervous system—neurons—are highly specialized and fairly fragile. When a person has a stroke, there is a very short window within which the loss of blood supply to brain tissue can be restored without permanent cell death.

Another problem is that the physical connections of the neuroanatomical structure of the brain do not at all translate to thought, emotion, or a sense of self. Neurons create and release neurotransmitters that respond to them via receptors that increase or decrease in number depending on a multitude of factors. They also create proteins and are affected by a vast number of local and longer-distance connections that affect mood, memories, and thought processes. The cryonics' view of freezing the brain to re-create that connectome is the equivalent of saying, *Let's re-create Spain. I have a detailed map. That should suffice.* The culture and the people are not embedded in geography, just as a sense of self and existing memories are not simply found in the neuronal mapping of the human brain (a feat that is itself still far from being achieved).

You might think, "Well, we don't know what scientific breakthroughs are possible in the future, so why not try cryonics?" There are a lot of reasons, actually.

First, let's talk about how cryonics actually works. We need to discuss why the process itself is fraught with problems, beginning with one pesky little molecule:

Water.

H_2O is a strange and beautiful molecule. Most every substance in the world shrinks when it encounters freezing temperatures, their atoms and molecules usually aligning in a compact, crystalline structure as their density increases. Not water. Everyone knows that ice floats. And not because it's riddled with bubbles, but because water volume expands when it freezes, and its density decreases in comparison to liquid water. Water molecules rearrange themselves not into tight, tidy crystalline structures, but into a looser structure with the molecules spaced out, due to the hydrogen bonding between oxygen and hydrogen atoms of different molecules. Many people have unfortunately come face-to-face with the realities of this expansion property when their house pipes burst during a deep freeze. Or when a soda can explodes in their freezer, peeling back the aluminum can in the process.

Because water makes up about 60 percent of the human body, and that water is located throughout the body, including inside the very delicate neural cells of the brain, when you freeze

Working with liquid nitrogen, which is used in cryopreservation

a brain, or a head or an entire body for that matter—havoc ensues. Those delicate neurons pop like little cans of Coca-Cola in your freezer. We're not talking brains exploding here, but on a microscopic cellular level, the neurons are being destroyed. Which is not a great situation if you plan on using those neurons again someday in the future.

Osmotic shock is another issue. As water begins to crystallize, the solutes dissolved within your cells—sugars, salts, and other chemicals—will concentrate outside the freezing water, and that can cause water to rush in to equal out the concentration, which can stress and kill cells. Some living things, such as certain frogs and tardigrades, those cute microscopic organisms colloquially known as "water bears," have figured out how to survive the freezing process. But they are not humans—they don't work the same way as we do. Which means those frozen bodies and heads in those Dewars might look outwardly preserved, but on a microscopic level likely have a lot of cellular damage.

All this freezing damage to human cells is a problem the scientific community has been tackling for years, with some good progress made. For example, human eggs, sperm, very young embryos, blood, and stem cells can be frozen (sometimes for decades) and still remain viable. But these are small amounts of cells, not entire organs like brains and hearts.

Not to mention whole bodies.

One method to avoid freezing damage is a process called vitrification. It is a method of freezing in which cryoprotectants are used to prevent ice damage to cells. They vitrify, or become glass-like without the expansion of frozen water. But often, delicate blood vessels in the brain collapse before cryoprotectants can fully perfuse the entire organ. Furthermore, cryoprotectants like dimethyl sulfoxide are often toxic to human cells. So the cells might get preserved, but will they live again? Not likely.

Nevertheless, the first human to be preserved posthumously via vitrification was a man named FM-2030. (No, this is not the start of a science fiction novel.) Born Fereidoun M. Esfandiary, he was a Belgian-born Iranian American who was not only an Olympian in the 1940s but a teacher and a philosopher. He also shook off the cultural trappings of conventions like first and last names, hence the unusual moniker. FM-2030 envisioned a future where, by 2030, people might live agelessly and with immortality—

a belief that made him what's called a "transhumanist," or someone who believes that technology should be used to enhance human bodies and cognition. Unfortunately, he succumbed to pancreatic cancer ("a stupid, dumb, wretched organ," he was quoted as saying).

Freezing technology aside, the brain is irreversibly damaged once it has been without oxygen for four to six minutes at room temperature. Furthermore, it's made up of several different parts, with different physical characteristics. The outer cortex of the brain has a different texture and biological makeup than, say, the corpus callosum, which is a bundle of millions of cells that helps connect the two hemispheres of the brain. It stands to reason that each part requires different methods for perfect cryopreservation. Despite the best attempts, organs can crack when freezing. Cells can burst. Fixatives or preservatives like aldehydes are toxic to tissues. They may preserve them, all right, but more like a preserved fetal pig on a shelf in a biology class. And no one wants that kind of preservative flowing through their veins or anywhere near their own functioning brain.

The man who called himself FM-2030 has a contemplative moment.

Not for the Faint of Heart . . . or Light of Funds

As if the science of cryopreservation wasn't hard enough to overcome, it's also expensive. In the 1960s and '70s, when cryonics began to draw public interest, most companies went out of business, and the corpses they were storing thawed and were subsequently buried or cremated. The US company holding Ted Williams's corpse, Alcor, charges $120,000 to have a body prepared for and kept in cryopreservation. If that's too steep, and if you're betting that your aged or ill body is unnecessary for life in the future, then

Billionaire founders of PayPal, Luke Nosek, left, and Peter Thiel have reportedly voiced their interest in a very cold future—being cryonically preserved after death.

a more affordable $40,000 will keep your head frozen for "neuropreservation" in its Dewars, which are all located in a nondescript building in a business district of Phoenix. (The Dewars are quite large and hold about a dozen human bodies, or three dozen heads. Plus or minus those pets.)

One big financial flaw with these companies is that after a person dies and is preserved, and the one-time fee (usually covered by a life insurance payout) is paid for freezing and storing a body, that's the end of the incoming money from that client. This means that companies can eventually fizzle out if demand wanes, thus the means to keep people frozen also fizzles out. However, where there were only two active cryonics companies a decade ago—Alcor and KrioRus (in Russia)—newer ones have since popped up in Michigan; Sydney, Australia; and Berlin, Germany. And the business is getting competitive. KrioRus has preserved 80 people and 50 pets and currently has 500 people waiting for the same treatment, but for roughly half the price of the US's Alcor. Which is to say that KrioRus is the Costco of human cryonics.

Then there's a whole other question that no one seems eager to answer. Who's going to pay for the cost of reanimation? The reanimated person? Maybe, if they have a flush trust fund for this. If not, they might awaken to an enormous financial debt, which could take another lifetime to pay off.

There are also legal issues to consider. Because the preservation process cannot begin until the person has officially become a corpse, cryonics lies in the realm of posthumous wishes about how to handle a deceased body. Some countries, including France, outlaw cryopreservation. Who votes on these things? Certainly not frozen dead people—or, rather, clients in biostasis. Cryonicists are careful about language, not referring to their charges as "dead," but "past their first life cycle." The lifeless bodies are "clients" and not "frozen" but "in biostasis" or "suspended." Revival is referred to as "reanimation."

Ultimately, the big question remains: If we bring frozen Bob back to life, either via a connectome uploaded somehow or through repaired DNA and successful defrosting—will it be . . . *Bob*? Proponents of cryonics have spoken to people who suffered near-drowning in frigid temperatures for over an hour, then woke up and were "themselves" afterward. But after hundreds . . . or *thousands* . . . of years, that person waking up might not be themself. If you believe the theory of a connectome being created digitally can happen, then that existence would be outside of the original person. A second, separate existence with no true connection to the life lived now. In which case, what were they saving themselves for?

Finally, there's another admittedly pessimistic thing to consider. When a person decides they want to be preserved for a future existence, it's with the assumption that the future will be a pleasant place in which to exist anew. That, for argument's sake, global warming won't have cooked and flooded the world into a barely livable place. Cryonics exists in a realm of speculative hope about a future that can revive people. But that speculation must also include the possibility that the world won't want to revive them so they can blissfully enjoy a new life. What if they were brought back to life in order to be enslaved, or were trapped inside a box to be experimented on?

Luckily, this horrific future is quite unlikely and won't happen, because cryonics is pseudoscience. It just doesn't work. People who are now cryogenically frozen are placing a bet on the future, a one-sided and glaringly optimistic one. In truth, they're really just paying for a very expensive way to take care of their corpses. But if it's the price they pay for facing death with a certain amount of peace, well, perhaps it's worth it. May they rest in peace forever—even if it's very, very cold in there.

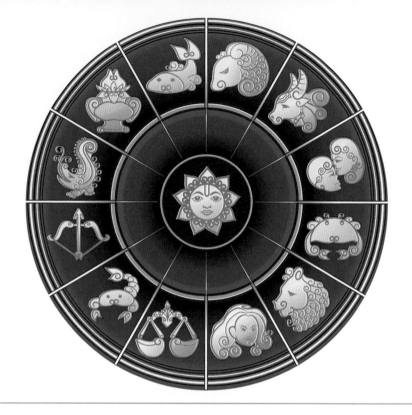

Astrology

On March 30, 1981, President Ronald Reagan left a luncheon speaking engagement at the Washington, DC, Hilton Hotel accompanied by his press secretary, James Brady, and his usual entourage of secret service agents. As Reagan headed toward his limousine, he passed by a group of press and citizens who had all been prescreened by security, except for one man: John Hinkley Jr. Somehow, Hinkley managed to elude security and penetrate the group. When the president was only a scant 15 feet away, Hinkley saw his chance. He crouched, firing his .22 pistol six times in less than two seconds, hitting Press Secretary James Brady, a police officer, and a Secret Service agent. Miraculously, he did not hit his target directly. The sixth bullet ricocheted off of the armored limousine and struck President Reagan, puncturing his lung and lodging in his body less than an inch from his heart.

Reagan's brush with death shattered his wife Nancy's world, which revolved around her husband. Nancy Reagan would later tell *Parade Magazine*, "It's a particular kind of trauma that never leaves once you've known it."

So when Nancy's Hollywood friend and television host Merv Griffin called her to say that he knew of an astrologer, Joan Quigley, who had predicted March 30 would be an unlucky day for her husband, it floored the first lady. According to the biography of Nancy Reagan by Karen Tumulty, the revelation was enormous.

"Oh my God . . . I could have stopped it!" Nancy exclaimed. She called Quigley immediately.

Even with the top security of one of the most powerful countries in the world, Nancy Reagan realized that her husband wasn't safe. But with Quigley's help, along with the guidance of the stars and the planets, perhaps she could protect her beloved in a way that dozens of secret service agents could not?

Quietly, Nancy paid Quigley thousands of dollars a month to choose dates and times for travel and events, including when to hold the State of the Union address. Quigley's involvement made the president's schedule chaotic and filled with odd times, such as Air Force One taking off at 2:11 a.m., and it also occasionally affected more pressing matters, like the president's cancer surgery. Shockingly, the directives weren't always just about timing. Prior to a 1985 summit with Soviet leader Mikhail Gorbachev, Quigley offered more than the usual scheduling advice, as she recalled in her memoir, *What Does Joan Say?*. "Ronnie's 'evil empire' attitude has to go," she told Nancy.

Chaos ensues after the assassination attempt on President Ronald Reagan outside of the Washington, DC, Hilton Hotel on March 30, 1981.

Joan Quigley was paid thousands for astrological readings that influenced the inner workings of the White House.

"Gorbachev's Aquarian planet is in such harmony with Ronnie's, you'll see . . . They'll share a vision." The White House tried to keep the astrologist's recommendations and bizarre scheduling instructions a secret for as long as they could.

Bizarre or not, did Quigley really tap into something most mere mortals don't know? Can the stars and planets steer us away from dangerous fates, or toward wealth, a soulmate, or even a good day to buy a lottery ticket?

Humans have endeavored to control our destiny for as long as we have lived. When life is mercilessly short (far shorter in the past than now), we do whatever we can to avoid catastrophe. Through the work of philosophers and scientists, including Hippocrates, Ptolemy, and Galen, humans often saw the microcosm of our bodies reflected in the macrocosm of the universe. So it was inevitable that the stars, moon, sun, and planets would work their way into our understanding of our own fate.

The Sunniest of Signs

The word "astrology" comes from the Greek "astron," or star, and "ology," or study of. "Astronomy" sounds a lot like it, but with the different suffix of either "onomia" for law or "nomos" for culture, meaning the "law or culture of the stars." And while astrology and astronomy are radically different now, from the time of Hippocrates to the Middle Ages, they were studied similarly and were sometimes considered interchangeable. Serious scholars studied both, along with alchemy, meteorology, and medicine.

Astrology as we know it today is the practice of understanding how celestial bodies affect people's experiences—a predictive art meant to advise us in our daily lives. Not until the late tenth or early eleventh century did Persian scholar and scientist al-Bīrūnī enumerate the differences between the two. Astronomy, he explained, was an academic endeavor aimed at understanding planets, stars, moons, their patterns and relationships to each other, and also their makeup. Astronomy was based in math and science.

Astrology was not.

Astrology was dedicated to discerning meaning among heavenly bodies and Earth as it pertains to human lives. But meaning and effect are not the same thing. Astrology once had two very distinct sides: "judicial astrology" (whether stellar bodies can foretell the future for a person) and "natural astrology" (the effect of the moon and sun on a person's health, for example). To be clear, planets and stars are too far away for humans on Earth to feel their direct electromagnetic or gravitational effects. But though we don't refer to it as natural astrology anymore, we do recognize that the moon and sun play a direct part in our health. Lunar cycles can affect sleep cycles, rates of car crashes, crimes, and suicides. Multiple sclerosis appears to be more frequent in babies born in months when maternal vitamin D might be low, due to less sun exposure. A study of 1.7 million patients treated at New York Presbyterian Hospital revealed some startling associations, such as higher rates of atrial fibrillation and congestive heart failure in those born in March, or ADHD peaking in those born in November. However, the study covered only one city, one environment. It's notable that asthma associations are linked to peak sunlight in New York in July through October. In Denmark and Austria, it's May through August.

Al-Bīrūnī (973–1050), Iranian scholar and polymath, separated the concepts of astrology from astronomy, which got him immortalized on a Russian postage stamp.

Sagittarius, the archer—from a certain point of view

But again, this is natural astrology. It is not the same as checking your horoscope in a magazine or on an iPhone app and trying to divine if you will fall in love that day. Or not.

Who hasn't looked up their zodiac sign to discern whether they're a true Libra, who, according to *Cosmopolitan* magazine, is prone to "adore beautiful things and culture, and seek perfect balance and harmony in all areas of their life"? Or sat down at a Chinese restaurant to learn from your scarlet paper placemat that being born in the year of the pig means that you're generally self-disciplined, smart, curious for knowledge, and calm? It all sounds great, but is it accurate? And more important, why do we care?

Before we talk about accuracy, let's get some definitions squared away. The Western astrological zodiac consists of signs that each correspond to one-twelfth of a belt of sky. In the second millennium BCE, Babylonian astrologers associated these signs with the constellations that the sun passed by as it rose and fell every day. These constellations and signs were Aries, Taurus, Gemini, Cancer, Leo, Virgo, Libra, Scorpio, Sagittarius, Capricorn, Aquarius, and Pisces. China used the entire sky, rather than just a belt of it, to create its own zodiac. Timewise, the dates corresponding to each sign in that zodiac begin with the March vernal equinox.

Some might think it absurd that one-twelfth of the Earth's human population all have exactly the same personality attributes or fortunes based upon these birth dates, or your "sun sign." But things can get far more specific than that through using a horoscope, which is Greek for "observer of the birth hour." A horoscope is a map of the sky at a precise time and place on Earth at its center, with each of the zodiac signs occupying one-twelfth of this circular chart, like 30-degree pie wedges. At a particular moment, the horoscope supposedly discerns how you might best handle your affairs, or what external events might occur. At birth, your horoscope is called a natal chart. The locations of the moon, sun, and planets at that precise moment are recorded within the pie. It might include several things, such as a person's sun sign, the part of the sky where the sun passed at the time of birth. An ascendant sign is located on the eastern horizon at birth. Lines connecting the planets and celestial objects are meaningful. Red lines mean they are in opposition, and blue lines mean they are synergistic. The angles of the connecting lines with respect to our view from Earth are called aspects and bear much weight.

The signs are further assigned elemental motifs, such as water, fire, earth, and air, as well as the properties of being cardinal, fixed, or mutable, depending on whether the zodiac sign begins, spans, or ends a season. Planets residing in certain elemental zodiacs impart their effects. For example, Mercury (being the planet of intellect, communication, memory, and transportation) residing in Libra gifts a person with an intellectual and justice-loving mind. Mars represents drive and courage; Venus, of course, exercises influence over love and the arts. Finally, 12 houses superimpose another 12-piece pie (they overlap greatly with the zodiac signs) that begin in your ascendant sign, and they signify things like appearance, finances, health, relationships, and cross-cultural relations and travel—all of which can come in handy for presidential scheduling.

Daily or monthly horoscopes are created by looking at the celestial events occurring at that time, the lunar calendar, and how these events play out in relation to the sun signs. But for many serious astrologists or astrology lovers, this is all just skimming the surface. A printout of a sample internet natal chart was 17 pages long, most of which was blurred out unless we paid $27.99 for the full horoscope.

Astrology is hugely popular now and has been for decades. Daily, weekly, and monthly astrology columns commonly run in respectable magazines and newspapers across the country, such as the *Washington Post*, *New York Magazine*, and the *Chicago-Sun Times*. Countless astrology apps are available. Approximately 70 million Americans read their horoscope daily, according to the American Association of Astrologers. A YouGov poll from 2022 showed that 27 percent of respondents believed in astrology. That number goes up to well over a third (to 37 percent) if you're under the age of 30. Half of Americans say they firmly don't believe in astrology, but that leaves nearly half who do—or aren't sure. In a 2012 General Social Survey, 34 percent of Americans believed astrology to be "very" or at least "sort of" scientific. That's quite a hefty number of people who believe that the

A Chinese illustration of Rahu, the Hindu Snake Demon who can swallow the sun or the moon, causing eclipses

positions of the stars and planets are playing with our fates on a daily basis, and that it's based in science.

Many cultures have astrology even more deeply ingrained into daily life. In India there are premiere astrology schools, and the celestial divination is enmeshed into many daily lives. In fact, astrology coexists very comfortably alongside science in India. Modern-day Chinese astrology takes into account all the birth statistics of a person to help predict auspicious personal relationships or marriages. In East Asian cultures, there is a sexegenary calendar cycle, or "ganzhi," which is a 60-year cycle of terms including "metal snake" and "earth rabbit." During the "fire horse" year of 1966, it was foretold that women born then would have a bad temperament and would kill their husbands. The fertility rate in Japan dropped from about 2.0 to 1.6 in 1966, most certainly to avoid the inauspicious birth year.

Science: The Enemy of Astrology

But how can astrology be accurate? What is the mechanism by which we can predict how to act, who to marry, based on where the sun, moon, and planets were at the time of our birth?

The simple answer is that there's no mechanism and astrology isn't accurate. As with any scientific theory, there has to be a mechanism by which everything on a natal chart, for example, acts upon human lives. And not only is there no good mechanism in astrology, but highly trained and esteemed astrologers can't agree on how these aspects can apply to the biology and physics of human existence. (They also have trouble agreeing on what should be on a natal chart and how it should be interpreted.) Because, again, the constellations and planets exert negligible amounts of electromagnetism and gravitational pull on people at that distance.

Furthermore, there are issues with the premise of the zodiac and birth charts. Have you ever heard of Ophiuchus? Probably not. Remember how the zodiac includes the 12 constellations that the sun passes over in the ecliptic? Well, a thirteenth constellation, Ophiuchus "the serpent bearer," wasn't included by the Babylonians for obscure reasons. Every few years, a few people stir up a kerfuffle about NASA being somehow responsible for deciding that the thirteenth zodiac sign now exists. People panic over the idea that if

An eighteenth-century engraving depicting the oft-forgotten constellation Ophiuchus, riding upon the far more popular Scorpio

you were born between November 29 and December 17, you'll no longer be a Sagittarius, but an Ophiuchus. However, NASA repeatedly likes to remind people that they had nothing to do with it. Ophiuchus was there the whole time.

Furthermore, because the constellations are different sizes (from our earth-bound perspective), the sun doesn't spend the same amount of time passing over each one. It spends 45 days passing over Virgo but is in Scorpio for only 7 days. Which points to the arbitrariness of the zodiac, for starters. And speaking of arbitrary, remember that constellations aren't flat pictures in the sky. Stars exist in three dimensions. The constellations that we assigned to our flat two-dimensional perception are but another example of the typical human habit of making pictures out of chaos. In other words, the constellations are somewhat meaningless—they're just flat line drawings we put together for our own benefit—but out in the universe from any other perspective, the picture falls apart. Also, what about the fact that Uranus, Pluto, and Neptune were added onto horoscopes after their discovery? And then what about after Pluto was demoted to a dwarf planet?

Many astrologers believe in the firm, unchangeable patterns in the sky and celestial bodies, but Pluto's discovery and subsequent demotion pokes a pretty big hole in this. Furthermore, many astrologers don't recognize that the zodiac positions designated by the Babylonians in the second millennium BCE in the night sky are actually different now. A famous astrologer, Elizabeth Teissier, once said that "the sun ends up in the same place in the sky on the same date each year," as an indication of the reliability of her art. But she was completely wrong.

According to astronomers, our view of the zodiac has been changing since it was unveiled three millennia ago. The Earth's rotational axis wobbles. As it spins, the sides of the Earth bulge outward a little (the way your hair flies outward if you spin around). The resultant gravitational pull by the moon and sun causes the Earth to wobble as it spins. So the Earth's axis rotates around in a circle over the course of 25,800 years like an unstable spinning top—a process called precession. This means that a sun sign for a person born on a particular date differs over time. The vernal equinox moved from Aries to Pisces from ancient times until about 100 BCE. In the year 2700, it'll be in Aquarius.

And finally, according to natal charts, twins (identical or fraternal) and babies born around the same time and place ought to share the same horoscope for their entire lives. But their lives are anything but 100 percent identical.

But perhaps that's not enough to deter believers. Many followers of astrology make a sharp demarcation between the fluffy, daily social media horoscopes decorated in pink and celestial decor and the serious astrologers who are renowned for being "scientific" in their work. But how accurate are the true experts? Can their work be critically analyzed?

In 1985, scientists tried to do so. *Nature* published a study by Shawn Carlson, a physicist, who tried to comprehensively test whether renowned astrologers who were considered high-quality could reliably gather true information about their clients based upon their natal charts. Carlson welcomed both the scientific and astrology communities into the design of the study to eliminate his own bias as a scientist. Twenty-eight highly esteemed astrologers from around the world were invited to the study, which was double-blinded. Volunteers offered their personal information, and natal charts were created for them by the astrologers. Then, the volunteers were given three natal charts—one that was their own and two that were not. Their challenge was to guess which was their own. Astrologers were hoping they would pick their own charts more than half the time. The result? The volunteers chose their own natal charts only 33 percent of the time—the equivalent of them randomly picking one of the three charts.

University of Saskatchewan researcher Ivan Kelly embarked on another study, in which he followed 2,000 people under the same zodiac sign and born within minutes of one another for 20 years. According to their birth information and astrological rules they ought to have had very similar traits, but they didn't. French psychologist, statistician, and astrologer Michael Gauquelin invited 150 readers for a free astrological reading. Ninety-four percent of them thought the reading was accurate. It turned out that the same reading was given to all 150 readers, belonging to the horoscope of a murderer who had killed people fleeing persecution from the Nazis.

And let's not forget all the psychological reasons why astrology is so attractive and believable. It's all too easy to selectively remember portents that come true and forget the ones that don't. Who hasn't looked at their star sign description and paid attention to only the attributes that fit? Then there is the Barnum effect, which is when you read a very general set of predictions or personality traits that seem specific to you, but really aren't. How appropriate that the effect was named after P. T. Barnum, the American showman famous for taking advantage of the public's gullibility—he coined the phrase "There's a sucker born every minute." Or, as Theodor Adorno wrote in his critique of popular occultism and astrology, they are the "metaphysics of the dopes."

In the case of Nancy Reagan, astrology represented control in a world that violently threatened the existence of the person at the center of her universe, who also happened to be the de facto leader of the free world. And while you may think Nancy Reagan and Joan Quigley were completely delusional about their beliefs in astrology, one fact is not in dispute: President Reagan never suffered another assassination attempt. Confirmation bias likely cinched Nancy's belief that she was doing the right thing, regardless of the millions of dollars spent on the secret service detail constantly orbiting them.

Astrology will continue to show up in Instagram bios, articles, and people's phone apps, giving them guidance on how to wrest control of their lives in a world full of uncertainty, loneliness, and danger. Perhaps a fate written in the sky is far easier to accept than one determined by our socioeconomic status, race, education, and the health foisted upon us by a world bombarding us with a vortex of good and bad choices. Astrology is perhaps a way of bringing order to chaos. Its advice can be cherry-picked at one's beckoning and according to one's need. Also, it's a heck ton of fun.

Ley Lines

While driving through Blackwardine, an English village and archaeological site in Herefordshire, Alfred Watkins had a uniquely compelling vision. It was a June day in 1921 and the professional photographer, businessman, and passionate antiquary was pondering an old Ordnance Survey map. In his mid-60s with a close-cropped gray beard, spectacles, and close-fitting suit, Watkins stood out among the green rolling hills and peaceful orchards of the Herefordshire countryside. Looking across those hills, he abruptly realized that he could draw a straight line on the map connecting points of ancient sites and landmarks, such as a Roman camp and a Bronze Age hilltop fortification. (If this "revelation" gave you pause, you're not alone. Because can't you draw a line connecting any random points on a map?) Watkins concluded the straight line must have been an ancient trackway—and that prehistoric populations of Britain had navigated by these straight lines and purposefully built significant structures along them.

Alfred Watkins

Watkins's son Allen later recalled, "Then without any warning it happened suddenly. His mind was flooded with a rush of images forming one coherent plan. The scales fell from his eyes and he saw that over many long years of prehistory, all trackways were in straight lines marked out by experts on a sighting system."

Watkins's theory was that Britain was crisscrossed by a network of straight alignments that connected ancient landmarks, such as standing stones, cairns, Roman camps, Bronze Age fortresses, and medieval castles. These sites of historical and antiquarian interest were connected by footpaths called "leys," remnants of which survived into the 1920s, which Watkins thought also served as travel and trade routes for historical peoples populating Britain before the arrival of Romans in 43 CE. Many of these leys followed along high ground, staying on ridgelines whenever possible. To Watkins, these seemed like sensible paths for ancient people to travel, a route where you could maintain clear sightlines to the countryside around you and best protect yourself from danger.

Watkins's curious and wide-ranging intellect was excited by the new theory. He presented it for the first time at a meeting of the Woolhope Naturalist and Field Club of Hereford in September 1921, expounded on the theory in his first publication, *Early British Trackways* (1922), and then codified it in his seminal work on ley lines, *The Old Straight Track*, published in 1925 when Watkins was 70 years old. *The Old Straight Track* remains in print 100 years later, despite its meandering points and obscure language, attesting to our continued fascination with the deep past.

As the Crow Flies

The term "ley" was used by Watkins to describe the straight alignments he sought out in the British countryside. The term itself, derived from an Old English word for "cleared space," really has nothing to do with roadways or tracks. However, Watkins convinced himself of the historical usage of the term in place names he noticed along his alignments, such as the Ley Farms, Wyastone Leys, and Tumpey Lay.

Carried away on the wings of pseudoscientific discovery, he even became convinced that the term "ley" was embedded in the evolution of the English language. "It was once absolutely necessary to keep 'straight on' in the ley, for, if you did not, you would be *de-leyed* on your journey."

You get the point.

As his enthusiasm for the theory grew, his claims gradually became more grandiose. Watkins invested the route surveyors, the original "ley men," with mythic spiritual importance to ancient peoples. These ley men, who used twin poles to set routes through the landscape, were, according to Watkins, viewed as seers by their contemporaries.

If you're curious what a ley man might have looked like, Watkins suggests you take a gander at the Long Man of Wilmington, a hill figure in East Sussex with murky origins. Hill figures are a type of geoglyph, or landscape art designed to be viewed from a distance. The English countryside is

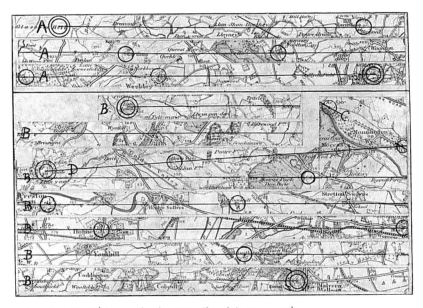

Ley lines in action! (aka, randomly connecting dots on a map)

The Ridgeway

Of course, some ancient British trackways really were set along ridgelines, the oldest and most famous of which is known today as "The Ridgeway." The route follows the chalk ridge of the Berkshire Downs from Wiltshire to the River Thames. In use for at least 5,000 years, the Ridgeway served primarily as a trading route connecting the Dorset coast and Norfolk. The high ground was indeed useful for travel, as it kept your feet dry (relatively speaking, of course—this is England, after all). The elevation also provided commanding views of the surrounding countryside, so you could scan for potential threats. During the Iron Age, hill forts were built along the Ridgeway to help protect trade, demonstrating its importance. A series of English invaders also utilized the Ridgeway to move armies through the landscape, including the Romans, then later the Saxons and the Vikings. In the medieval era, drovers traveled it as well to drive cattle to market towns. Walking the pathway today—an 87-mile stretch of which is preserved in the UK as a National Trail—really does provide an opportunity to physically connect with centuries of English history.

Those who took the time to engage with the theory criticized it in several key areas:

1. If you throw a stone in England, you'll hit a site of historical or antiquarian interest. It's not too difficult to connect them with lines on a map.

2. Straight lines are not always a sensible way to travel through the landscape and would have sometimes been wildly impractical as a method of traversing hilly and mountainous landscapes.

3. Watkins's leys connect points from significantly different historical eras, weakening the case that they were purposefully aligned at some point in prehistory. For example, a neolithic long barrow and an Iron Age hill fort

particularly famous for these hill figures, often cut into grass to reveal the starkly white chalk bedrock underneath. The Long Man of Wilmington, stretching 235 feet tall, depicts a masculine figure holding two staves, one in each hand.

could be sighted along the same ley line; however, those two constructions might be separated by as much as 4,000 years. It's sort of like sighting Stonehenge with Victoria Station and assuming they were intentionally built along the same line.

4. Too many of the alignments weren't really all that straight after all and required approximation to conform with the theory.

Regardless, the idea of "leys" captured the imagination of amateur archaeologists looking for intellectually stimulating excuses to get out into the landscape. Watkins helped form the Straight Track Club in 1926, and hunting for leys became a hobby for many people with a desire to explore the outdoors while searching for clues. This was not dissimilar from a hobby like geocaching today, where participants utilize a GPS device to hide and seek containers called "geocaches," hidden all over the world. To help these ley hunters in their tasks, Watkins even published a *Ley Hunter's Manual*, which they could tuck into their rucksacks on their outings.

The Straight Track Club's heyday lasted from the late 1920s until the mid-1930s. Watkins, the originator and eternal champion of the idea, died in 1935, and while the Straight Track Club continued on for a few years, the looming outbreak of World War II brought an end to such leisure pursuits.

At this point, the concept of ley lines would likely have faded into obscurity, were it not given an entirely new life in the Age of Aquarius.

Questionable advice for rogue amateur archaeologists aside, Watkins's ideas were relatively harmless. They were also uniformly ignored or rejected by professional archaeologists of the 1920s.

The Nazca Lines

Unlike the ley lines of England, which require a large map and an even larger imagination to uncover, the Nazca Lines, deep in southern Peru, are readily visible to anyone. Over 800 straight lines, as well as 300 geometric figures and 70 animal and plant designs crisscross the desert landscape. Constructed by people of the Nazca culture at least 2,000 years ago, if not more, the lines have intrigued and mystified people since they first entered the historical record in 1553. By that time, the Nazca culture had long since collapsed, disappearing around the year 500 CE and leaving no written records behind. The original meaning behind the lines has been lost to time.

Why did the Nazca culture build the lines? What was their intended purpose? It's a legitimate mystery, one that many scholars over the centuries have tried their hand at explaining. Without any evidence to back up a theory, it's anyone's guess. And plenty of guesses have been made: astronomical markers, religious practices, ceremonial purposes, calendar markings, irrigation, trackways.

As with all ancient mysteries, pseudoscientists have entered the fray and taken it a bit . . . further. Erich von Däniken, author of *Chariots of the Gods*, declared that the lines were intended as landing strips and navigation aids for UFOs. Instead of recognizing substantive Indigenous ability to modify the landscape, von Däniken declared that aliens themselves had built the lines to help them guide their spaceships back to a favorite part

Ancient Aliens, Chinese Dragons, and Transcendental Meditation

The rise of the countercultural movement in the 1960s also led to a rise in counterfactual thinking. When English writer, intellectual, and esoterist John Michell dusted off a copy of Watkins's *The Old Straight Track*, he embedded another layer of meaning on top of these supposedly ancient lines crisscrossing the English landscape. Where Watkins saw a network of roadways, Michell saw a network of guiding strips for UFOs.

of the Earth where the people worshipped them as gods. Despite being summarily dismissed by all serious scientists of Nazca culture, von Däniken found a ready and willing audience excited to believe his ancient astronauts' theory.

With the success of the book came a significant uptick in tourism to the Nazca lines in Peru, ironically causing negative environmental impacts to the fragile landscape that contain them.

Intriguing designs found in the Nazca lines

Michell subscribed to a belief in ancient astronauts, visitors from faraway planets who frequented Earth in the prehistoric era, where they were worshipped as gods. In Michell's view, these aliens eventually grew disgusted with the increasingly materialistic leanings of humankind and departed for the stars, permanently, as human technology advanced and greed for more and more possessions took hold.

The extraordinary irony of a species sufficiently advanced to have discovered interspace travel turning its back on another species for stumbling its way into the Bronze Age was not apparently debated.

John Michell plotting out energy conduit lines, while enjoying a beer

Michell's views were first espoused in *The Flying Saucer Vision* (1967) but were codified in *The View from Atlantis* (1969). The latter is described by British historian Ronald Hutton as "almost the founding document of the modern Earth Mysteries movement." In it, Michell turned East for an ancient source for his theory, connecting the idea of ley lines with the Chinese mythological concept of "dragon paths," or lung mei energy lines. Such paths were believed to conduct and carry energy across the Earth.

"It was recognized that certain powerful currents, lines of magnetism, run invisible over the whole surface of the Earth," wrote Michell. "The task of the geomancer was to detect these currents and interpret their influence on the land over which they passed. The magnetic force, known in China as the dragon current, is of two kinds, yin and yang, negative and positive, represented by the white tiger and the blue dragon. The lines of this force follow, for the most part, mountainous ridges and ranges of hills."

Without citing any Chinese sources as evidence, Michell added: "The striking beauty and harmony of China, which all travelers have remarked, was not produced by chance. Every feature was contrived."

According to Michell, the Chinese dragon lines were utilized by the "native geomancers" to direct the flow of magnetic current and "fertilize the countryside." So if you had a particularly good crop one year, you could thank your friendly local Chinese geomancer for impregnating the land on your behalf.

It wasn't much of a leap from here for Michell to apply the same concept to the British countryside and the "old straight tracks" that ran along ridges and hill lines. He declared the ancient sites and monuments connected by ley lines in Watkins's theories were places where natural energy forces of the Earth were concentrated. Energy flowed between these sites along these straight tracks and, for the adept mystic, presented an opportunity to tap into this mysterious energy force.

Ancient aliens, who used ley lines to help land their spaceships

What that energy entailed or what could be done with it, if harvested, was left to the reader's imagination. And this flexibility allowed for a wide range of interpretation as the New Age movement came to life in the early 1970s.

The idea of ley lines serving as energy conduits fueled a resurgence in interest in the old theory. Ley hunting clubs sprang up again in Britain. The timing was brilliant: Thanks to the rise of New Age thought, Wicca, and neo-paganism, there was an appealing audience willing to believe in a prehistoric Britain guided by seers manipulating energy forces along straight lines in the earth. A magazine called *The Ley Hunter* began publication in 1969 to support the resurgence in interest.

A line of prehistoric stones, purposefully built along a ley line to, um, maximize energy conduits.

Various ley theorists expounded upon Michell's ideas until the general theory worked something like this: Prehistoric Britons had discovered in the landscape places of concentrated power, or energy, and constructed their temples and stone monuments on top of them. Gradually, these Britons realized that this natural energy in the landscape could be controlled and directed throughout the countryside through the construction of energy conduits, that is, ley lines. Stone avenues, monuments, temples, and barrows were all built along these energy lines, and once they were connected together through the leys, British geomancers were able to significantly amplify the power in the landscape and direct it toward practical ends, such as hypothetically producing four harvests in a given year.

While the advance of technological developments and modern construction had wiped out many of these natural avenues of energy, they could still be seen in the landscape and were embedded in Watkins's ideas of the "old straight tracks." For the modern-day ley hunter awash in the ideology of the New Age, discovering these places of potent "energy" had potential spiritual benefits. Heightened states of awareness and transcendental consciousness were potential, and potent, byproducts of walking a ley, or entering a barrow, or touching an ancient stone monument.

Laymen . . . or Ley Men?

While Watkins may have not conceded the point, his ancient surveyor "ley men" are not the origins of the modern-day word "layman." In common usage, the term "layman" denotes someone who is not an expert in a particular field. However, that meaning is secondary to the origin of the term, from the Greek "laikos," or "of the people." The term "layman" was first introduced into English to separate clergy from non-clergy. Indeed, *Merriam-Webster* simply defines layman as "a person who is not a member of the clergy." Thus, most of us are, in the original meaning of the word, laymen. Regrettably for the ley enthusiasts, the dictionary does not also include "ancient surveyors of the British landscape" as a third definition.

Enthusiasms aside, the problem with the ley-lines-as-energy-conduits theory was that it could never be effectively demonstrated or historically proven. Eventually the idea wore itself out and the interests of its supporters spread further afield into adjacent beliefs, such as ancient astronauts and archaeoastronomy. *The Ley Hunter* magazine ceased publication in 2005, its last editor calling the idea "dead."

But of course it never really died. The mantle is continually carried forward by esoteric enthusiasts, some of whom claim to have had profound spiritual experiences while exploring these ley lines. (And we aren't here to dismantle deeply felt religious experiences.) The historical evidence behind ley lines, however, so blatantly destroys the theory that these monuments were purposefully aligned by prehistoric seers as to leave no room for doubt.

To this day, no ley hunter has successfully been able to align a ley utilizing exclusively prehistoric or megalithic sites. All leys have been dependent on the inclusion of much later construction, sometimes thousands of years apart. Medieval churches, for example, are frequently included in ley lines despite having been built many centuries after the prehistoric sites to which they are supposedly connected. Further, medieval churches were built by highly literate people who left behind detailed writings of the decisions made in the construction of their churches. Not a single source mentions anything like a ley or a purposeful alignment with other structures in the landscape. The entire ley theory has zero evidence, not even a shred, in the historical record.

The alignments sighted by Watkins and other enthusiasts can be explained entirely by the alignment of random points. As statistics readily demonstrate, when a large number of random points are marked on a flat surface, straight alignments are a byproduct. And in England, the countryside is so deeply embedded with physical remnants of the past that random points are easily found. Random points theory was memorably demonstrated by the archaeologist Richard Atkinson, who marked the positions of telephone booth boxes on a map and drew straight lines between them, thus proving the existence of "telephone box leys." (Telephone companies as highly advanced geomancers, anyone?)

And what would Watkins think of how people took his theory and ran with it? Well, he would probably be a little disappointed with the way his theory was wildly exaggerated by the Earth Mysteries movement. Watkins was a rational, highly intellectual individual, active in an era when amateur archaeology still accounted for many archaeological advancements in the field. His enthusiasm for prehistory and knowledge of the British landscape may have carried him a little far with his theory, but he correctly identified some aspects of ancient British landscape management, such as large-scale prehistoric deforestation, that were later validated by professional archaeologists. He would likely have little patience with UFOs, or energy conduits, or other pseudoscientific elements that have become so intrinsically bound up with his theory.

At the end of the day, the ley line theory probably found such resonance despite its lack of evidence because it's . . . *fun*. Poring over old maps, seeking out connections between mysterious places from the far-distant past, and then walking the beautiful British countryside to connect them . . . there are far worse ways to spend a Saturday afternoon.

Personality Psychology

"Eat for 24 Hours Straight and Reveal Your 100% Career Match!"

"Which Dessert Would You Be in Another Life?"

"Which Star Wars *Character Matches Your Sleep Schedule?"*

The personality quiz. It feels ubiquitous on Facebook and BuzzFeed and has long lurked on the pages of *Cosmopolitan* and in teen magazines. But it actually goes back further still, to ladies' magazines dating to the end of the nineteenth century. In today's world there's a good chance quizzes have crossed your social media feed, where you've likely succumbed to the siren song of clickbait to test your knowledge of favorite snack foods in each state.

Personality type tests are irresistible for so many reasons. They claim to be a psychological X-ray, revealing your tender and terrible insides. Who doesn't want to know what kind of person they *truly* are? Wouldn't you want to belong to a club, especially if you've always felt like an outsider? This could be the way to find your perfect soulmate or understand why you keep sabotaging friendships without knowing it.

Personality tests deliciously unearth things about you that even *you* didn't know ("I'm a social introvert, not an anxious introvert? Wait, there's four kinds?"). Some questions on personality tests might be fairly superficial ("What is your favorite donut?"), but other tests might ask you something that seems more probing, like:

Do you avoid crowds?

Are you always prepared?

Do you care little about unhoused people?

Are you more inclined to follow your heart or your head?

At the end of a tiring week, would going to a big party be exactly what you needed?

"See, Jane? This personality test indicates that wearing your corset will ensure the space-time continuum will stay intact."

Just My Type!

Many personality tests aren't just silly bubblegum-pop quizzes. Scroll around Twitter or Instagram and chances are you'll notice many people's bios include a cryptic four-letter designation, like INFJ, ENFP, ESTJ, or another of the 16 personality types allotted by the now-famous Myers-Briggs Type Indicator (MBTI) test.

The MBTI is a questionnaire that prompts you to answer a long series of questions, like whether or not you're likely to get tasks done, are outgoing or shy, make friends easily, or notice other people's emotions. The theory is that people tend toward modes of how they gather information (Sensing or Intuition), how they assess the world (Judging or Perceiving), where they place their attention (Introversion or Extroversion), and how they analyze things (Thinking or Feeling). Each preference has a letter assigned to it, so there are 16 possible combinations, one of which you are assigned at the end of the test. And each four-letter combo is assigned a designation to give the alphabet jumble clarity, like the ENTP "Inventor," the INTP "Architect," the ISFJ "Protector," the INFP "Healer," the ISFP "Composer," or even the INTJ "Mastermind." Honestly, they all sound pretty great.

Perhaps a little too great.

People and companies love the MBTI—and the numbers show it. CCP, the company responsible for marketing the test and certifying trainers and the exclusive publisher of the Myers-Briggs Indicator assessment, brings in more than $20 million annually in profits and possesses a staggering 800 products related to the assessment. An estimated 2 million people take the test every year, and about 50 million have taken it since the 1960s. At least 10,000 companies use it (particularly after polygraph tests were outlawed for employment recruiting in 1988), as well as 200 government agencies and 2,500 colleges and universities. And it all started with two women, Katharine Cook Briggs and her daughter, Isabel Briggs Myers.

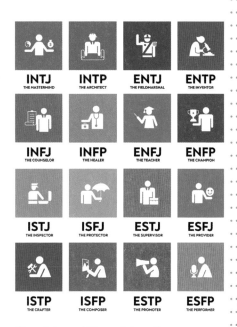

The rainbow of Myers-Briggs Type Indicator personalities

Katharine, born in 1875 to a distinguished science professor father and devout Christian mother, was a highly intelligent child who skipped grades and sported thick glasses and had a quiet demeanor that belied a prolific imagination. After college and marriage, Katharine threw herself into the science of child-rearing when her daughter Isabel was born in 1897. She would subject Isabel to "No! No!" drills, in which Isabel would be shown a candle flame, for example, then spanked and peppered with "No! No!" if she reached for it. Her attempts to understand personality began with trying

Photo of Katharine Briggs and her daughter, Isabel Briggs Myers, taken in the early 1900s

to suss out human development. She believed children needed "submission to necessary authority" and later as adults, "specialization." Later, her attention turned toward personality work in the creation of a sort of "people-sorting device."

Isabel was an accomplished child, learning several languages, practicing craft skills like metal work, and publishing six short stories in magazines by the time she was a teen. She attended Swarthmore College and married, after which her mother noted that her son-in-law, Clarence "Chief" Myers, had such a different personality from those of the Briggs family. The Briggses were intellectually curious and creative, "big thinkers." Chief seemed more detail oriented and practical. Katharine couldn't get a read on Chief. So, she began studying biographies and self-learning about psychology, formulating her own groupings of personality types. Isabel would later say that the reason she and her mother created the MBTI was simply "because I married Chief."

Katharine soon became rather enamored with the theories of Carl Jung. The Swiss psychologist and psychiatrist who was a contemporary (and for a period of time, a friend) of Sigmund Freud wrote a theory of personality types, *Psychologische Typen*, in 1923. Jung discusses two main types of personalities, introverted and extraverted (his spelling), as well as four "functions": sensing, intuition, thinking, and feeling. (Katharine also became obsessed with Jung himself, writing letters to him in abundance, creating lyrics to songs about Jung, and writing personal, private erotica featuring him.)

When Katharine came across Jung's book, years after Isabel married Chief, her ideas about personality types melded with his. Later, Isabel threw herself into the work as well, after World War II inspired many to be productive for the sake of their country. Her contribution in this realm led her to help people find their fitting vocations instead of going against the grain of their personalities. Isabel's daughter-in-law, Katharine Myers (one of the last holders of the MBTI copyright before it was passed on to CPP), noted that "Isabel decided . . . if she could give people access to knowing their psychological type, it would be a contribution to world peace." The test developed as time went on, at one point ballooning to a whopping 120 questions, before it became the 93-question MBTI that we know today.

Psychologists Say . . .

Given that MBTI has such a firm foothold in the corporate world (and everywhere, let's be honest), it's a bit shocking to learn that, for the most part, the professional and academic psychology and psychiatry establishment relegates it to the realm of pseudoscience.

The strict categories the test enforces took too many liberties with Jung's original ideas. Jung himself would never have gone this far. He even said "Every individual is an exception to the rule" and "To stick labels on people . . . was a childish parlor game." The MBTI also classifies you exclusively as introvert or extrovert, and once again, Jung had something to say about that, pre-MBTI: "There is no such thing as a pure extravert or a pure introvert. Such a man would be in the lunatic asylum."

Jung's thoughts on personality classification aside, contemporary social scientists and psychologists know that there exists a range of behaviors on a spectrum, and personality facts are seen as a continuum on that spectrum. The dichotomies of the MBTI are problematic for a few reasons. Though "introvert" and "extrovert" might be opposites on one spectrum, "judging" and "perceiving" are less linearly related, and the same is true for "intuiting" and "sensing." One might say that liking sweets and

Carl Jung, the subject of Katharine Briggs's studies and her erotic letters

liking savory foods are opposites, but when you think about it, they aren't. Not really.

Then there is the very positive, well, spin of the designations. The descriptions of the types are not just flattering, but a bit blurry as well. The Forer effect, also called the Barnum effect, is on full display when it comes to the MBTI. This occurs when you tend to think certain vague but generally positive attributes assigned to you are true. Consider the titles of the MBTI types: "Virtuoso" and "Performer" and "Adventurer." Not very scientific sounding, but quite flattering! Which calls into question the validity of the test as a legitimate psychological tool.

The numbers also test the MBTI's validity. How accurate is the test? How reliable? Some studies have shown that 50 percent of people who take the MBTI later pick an entirely different type. Professionals in the fields of psychology and psychiatry generally disregard the MBTI and don't use it as a part of their practice, citing the dearth of reliability in studies. Without test-retest reliability, you can't have validity in a personality test, as personalities are considered generally stable across lifespans. As such, the validity of the MBTI has been doubted for decades. The American Psychological Association notes on its website that it possesses "little credibility."

The MBTI is part of a corporate training industry that makes money hand over fist, ostensibly trying to help businesses hire the most ideal candidates for certain jobs or assist their employees in growing to be the best they can be. For example, you might use the MBTI to show how those in management positions might communicate well or poorly because of their personality types. And it's used widely by companies to screen new job candidates, which is quite problematic, given that you're supposed to hire people based specifically on their perceived ability to perform job-related tasks. The Equal Employment Opportunity Commission says that the hiring process should be "properly validated for the positions and purposes for which they are used." The MBTI has not been validated and proven predictive for the millions of specific job titles in the world.

The personality testing industry brings in $2 billion annually, so it's in the industry's best interests that any studies done on the MBTI or other tests are positive. It's no wonder then that the vast majority of MBTI studies are published in the company's own publication, *The Journal of*

Psychological Type. This is of course a glaring conflict of interest and a red flag (studies funded or sponsored by those who benefit directly from their outcome are an obvious no-no).

It's one thing for people to use the MBTI for themselves for fun or self-reflection. But what's frightening is that the test could easily be used as a nonvalid reason to fire people, limit promotions, and derail lives. There are serious consequences beyond just money. Right now, around 200 federal agencies still use the MBTI, including the CIA and the military. The Myers-Briggs website itself boasts that its products are "used widely within the federal civilian, state and local government sectors." That's American taxpayer money being employed for a test that has little credibility among research psychologists, according to the American Psychological Association.

Perhaps the bigger question is: Why do we care so much about personalities and personality types? Or maybe it's more existential, as in: What actually *is* a personality?

Humor Me!

The American Psychological Association defines a personality as the "enduring characteristics and behavior that comprise a person's unique adjustment to life, including major traits, interests, drives, values, self-concept, abilities, and emotional patterns." Boiled down, a personality is what makes you *you*. Imagine that a perfect replica of you showed up one day, with all your memories and mannerisms and was wearing clothes from your wardrobe—your personality is what would allow your friends, coworkers, and families to know instantly that it just wasn't *you*.

There's also a difference between personality *traits* and personality *types*. Traits are behavioral patterns that are consistent in a person; types are a collection of traits that make up a broader collective group. Generally, traits are consistent in a person over time and in different situations, they differ between people, and can be reliably measured either by an individual or by others. In the world of psychology/psychiatry, types aren't really used; traits are.

But types have been considered, defined, and used for countless years by non-psychologists. One of the earliest personality type theories was

based on the humoral theory. Variations of this theory hearken back to at least the time of ancient Greece with the writings of Hippocrates in 400 BCE, and even earlier versions appear in Ayurvedic writings. The concept was that the human body's four humors, or fluids—blood, black bile, yellow bile, and phlegm—were in a particular balance in the body. The humors were associated with the earthly elements of air (blood), earth (black bile), fire (yellow bile), and water (phlegm), respectively. Any imbalances were believed to cause illnesses.

In the second century CE, Greek philosopher and physician Galen further crystallized the relationship between humors, the earthly elements, seasons, and personalities, or temperaments. He wrote, "Sharpness and intelligence are caused by yellow bile in the soul, perseverance and consistency by the melancholic humor, and simplicity and naivety by blood." Writing in the nineteenth century in *On the Constitution of the Universe and of Man*, J. L. Ideler went further: "The people who have red blood are friendly [sanguine]. They joke and laugh about their bodies, and they are rose tinted, slightly red, and have pretty skin. The people who have yellow bile are bitter, short tempered, and daring [choleric]. They appear greenish and have yellow skin. The people who are composed of black bile are lazy, fearful, and sickly [melancholic]. They have black hair and black eyes. Those who have phlegm are low spirited, forgetful, and have white hair [phlegmatic]." (For the latter, it makes you wonder if maybe taking Mucinex would suddenly improve your mood and memory and revert your hair to the color of your youth.)

Contemporary understanding of human physiology led to the humor theory being tossed out well over 100 years ago. What's replaced it? Well, a lot of things. Take, for example, the idea of a choleric personality, with associated traits such as being active, restless, aggressive, impulsive, and changeable—all of which show up in a modern equivalent of the so-called type A personality. You know them. The perfectionists. The ambitious ones who work 24-7, who write 50 items in their bullet journals every morning at 5:00 a.m. The ones who look like they're going to have heart attacks from

LEFT: *The Hippocratic four humors (clockwise from top left: blood, yellow bile, black bile, and phlegm) and their associated personalities: sanguine (optimistic), choleric (irritable), melancholic (sad), and phlegmatic (calm).*

Body Type BS

The concept of body types—ectomorph (tall, thin, trouble gaining weight and muscle), mesomorph (neither gains nor loses weight easily), and endomorph (a tendency to be overweight with excess fat, difficulty losing weight)—has been a popular method of classifying human bodies among the public. But not among scientists. Further, it's been used to define how one ought to exercise and diet in very popular fitness schemes. The origins of these body types draw on a 1940s study by William H. Sheldon, who attempted to make a connection between personalities and body types. For example, an ectomorph tends to be shy, introverted, and thoughtful, whereas an endomorph is an extrovert who cries when stressed. Further, he used scandalous nude photos taken of incoming Ivy League and Seven Sisters college students to do his research. The concept of these three body types has been debunked, but the legacy lives on. Just do a web search for "body type" and you'll get nearly three billion hits, with the top results including these three terms.

stress fairly soon. You're no doubt familiar with these types and have probably classified yourself as either type A or B, the latter being someone who is more receptive, chill, and flexible.

Where did the type A and B personalities come from? Funny story. The concept caught on in the 1960s, and the research led many to believe that behaviors associated with the type A personality could increase your risk of heart attacks. If you could mitigate these behaviors (Calm down! Relax!), then maybe you could decrease your heart attack risk. Ads showed grinning, exuberant folks who were "calmed" by cigarettes. The problem was, during the 1960s and through the 1990s, many of these studies were funded by tobacco companies, which hoped to prove that it was often human behaviors responsible for heart attacks, not necessarily cigarettes. The companies Philip Morris (maker of Marlboro) and R. J. Reynolds (maker of Camel) collectively provided tens of millions of research dollars. The hope? That the medical research might sway the growing public health push against tobacco.

So *did* the studies show that the type A personality confers risk for heart attacks? Not really. Most of the research couldn't definitively prove the assertion. And part of the problem was deciding on what a type A personality really was. Another issue was that most of the positive studies that found a link were funded by tobacco companies, and hence were subject to considerable bias.

But the research wasn't all misleading. It turns out that two traits usually found under the type A personality umbrella—anger and hostility—are actually linked to about a 20 percent higher risk of heart disease. The upside to this research was that the field of health psychology became a real thing. Health psychology now tries to understand the very complex web of biological, social, emotional, and physical factors that influence health. And in the world of health psychology, a scientific model of personality does exist and has good science to back it up.

It's called the five-factor model of personality, aka the Big Five, known by the acronym OCEAN. There's Openness to experience (curious or cautious), Conscientiousness (efficient and organized or extravagant and careless), Extroversion (outgoing and energetic or reserved and solitary), Agreeableness (compassionate and friendly or rational and critical), and finally Neuroticism

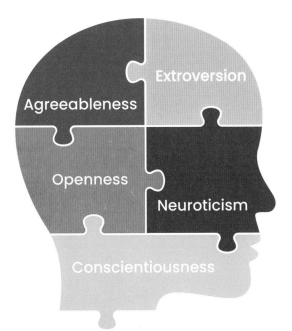

The elements of the Big Five personality model

(sensitive and nervous or resilient and confident). Unlike the MBTI, the Big Five is graded on a continuum. It's not a perfect test, but overall has been more reliable and has better predictability when it comes to job performance and even things like divorce and mortality.

Putting It to the Test

There is no shortage of personality tests out there, some good, others less so. Here's a smattering: ColorCode, CliftonStrengths, Insights Discovery, DiSC Model, Temperament Inventory, Hogan Personality Inventory, HEXACO, Unicru Personality Test, Enneagram, Socionics, Kiersey Temperament Sorter. Companies are using them to sort through massive numbers of applicants in search of good candidates. In fact, 88 percent of Fortune 500 companies use them. In the workplace, they are used as "team-building" exercises or ways to smooth communications between people. In job hiring, one round in the interview process might include an MBTI test. And troublingly, AI is being developed to use word choices in interviews to evaluate your personality in the hiring process.

But are personality tests prone to being racist, sexist, ableist, and biased against lower socioeconomic strata? Are they an invasion of privacy? Do they fail to take into account that a neurodivergent applicant could be a stellar job candidate, but is being eliminated from the running via a few personality screening questions? The answer to all of these is quite possibly yes. The MBTI has made it clear that their test ought not to be used for hiring purposes. But companies might use it, or other tests, anyway. Because they aren't considered "medical tests," they don't overtly violate the Americans with Disabilities Act (ADA). And so, based on the results of these tests, employees might be placed in jobs they don't enjoy, or be denied opportunities for upward mobility, or simply never be hired.

Even the Big Five, with a more robust scientific basis, is far from perfect. The test was developed in the US with the English language as a fundamental part of its analysis. It's also used in Europe, by countries that are considered WEIRD, or "Western Educated Industrialized Rich Democratic." But for countries that are non-WEIRD, the Big Five test isn't as reliable and can distort the true personality of the test taker. At times, it could be because the original language analysis used in the tests doesn't quite fit with concepts of personality across all cultures. For example, a person who grew up in rural Kenya may not be very conscientious about following time, compared to the punctuality-driven cultures in WEIRD countries. Sometimes it's due to a cultural tendency to say "agree" to questions and statements simply to be agreeable, which is called "acquiescence bias."

There's one more consideration when it comes to personality tests, scientific validity aside, and it's *where* you take them. We've all seen the stream of pop psychology quizzes that run through our social media feeds. They're hilarious and fun and occasionally insightful. But did you know they aren't as innocuous as you might think?

In 2008, a researcher named Michal Kosinski, then a graduate student at Cambridge University, created a personality quiz on Facebook that used similar questions from the Big Five. The information, if the user agreed to share their results with the researchers, showed how Facebook profiles related to these psychometric scores. The massive amount of psychometric data was used to create an algorithm that could more accurately predict how people might act or feel about . . . well, anything. If you had clicked "like" a minimum of 70 times while scrolling on Facebook, the algorithm could guess how you might think or act regarding voting for a political candidate or what movie you're likely to watch. It could guess your sexual orientation with 88 percent accuracy and your political party affiliation by 85 percent. It basically knew you better than your closest friends. If you had a minimum of 300 likes, the algorithm knew you better than your most intimate soulmate.

A fellow Cambridge researcher, Aleksandr Kogan, asked Kosinski about using this new database for a company, Strategic Communication Laboratories (SCL). It turned out that SCL was an election management agency that purported to influence elections all over the world. Though Kosinski turned him down, SCL somehow managed to create its own version of the same Facebook database. Both Kosinski and Kogan left Cambridge (Kosinski graduated with his PhD and

Cambridge Analytica whistleblower Christopher Wylie testifies before a Senate Judiciary Committee in Washington, DC, on May 16, 2018.

landed a job at Stanford; Kogan left for Singapore). But Kogan then created his own app called "This Is Your Digital Life" and amassed an enormous amount of Facebook data. By this point, SCL had spun off a new company that would hope to affect elections with more precision: Cambridge Analytica.

Though Cambridge Analytica denies using Kosinski's original work and isn't affiliated with any part of Cambridge University, it's interesting that the company chose that name given the parent research university. As for the company's role in Donald Trump's campaign victory in 2016, Cambridge Analytica CEO Alexander Nix said: "We did all the research, all the data, all the analytics, all the targeting, we ran all the digital campaign, the television campaign, and our data informed all the strategy." Regarding the usage of psychological targeting in politics, Michal Kosinski (now Dr. Kosinski) has said, "This is not my fault. I did not build the bomb. I only showed that it exists."

Like it or not, if you've been an avid social media user, you helped build the bomb, too. In any case, whether you subject yourself to an Enneagram, or take a quiz about what aquatic sea monster you are based on your social habits, or do a test to find out whether you have the chops to be a management leader in your company, one thing is for sure: People living in WEIRD countries can't get enough of themselves.

· · · · ·

No matter the cause—narcissism or boredom, earnest self-reflection or simmering ambition—personality tests will always be an attractive thing. We all love comfort ("I belong to a group!") and certainty ("You definitely belong HERE!"). Just know that falling for these tests without critically examining what they are is a mistake. But critical examination is not that fun. And these tests are fun. That's why they're so popular.

Remember also that while you're online, corporations are watching you and gathering every bit of data you share. Your significant other might not spill your secrets, but your laptop may. And be aware that the fun rapport-building test you're taking at work could be used to discriminate against you and ultimately fire you. Perhaps we should remember the wise words of Carl Jung, who said it best: "The world will ask you who you are, and if you don't know, the world will tell you."

Auguries

Romulus and Remus, the legendary twins who founded Rome, had a problem.

After a divine birth and a youth spent in exile, where they were famously suckled by wolves, the two young demigods were finally coming into their own. Upon learning about their royal heritage, the two brothers surfaced from the wilderness and set off to found a city together. They arrived between two hills in central Italy and agreed that the surrounding area was an ideal site for their new city.

But then things went south. Fast.

Romulus was convinced they should build the city on Palatine Hill. Remus, however, was not having it. The nearby Aventine Hill, he said, would be easier to fortify against attack and was clearly the better choice. Given that neither would-be urban planner anticipated that regardless of where they initially placed the city, even a slight amount of urban growth would quickly subsume both hills, they fell into disagreement. Arguments went back and forth between the two of them until they finally decided to seek guidance from the divine.

Agreeing to settle their dispute by augury, the Greco-Roman religious practice of interpreting the will of the gods through bird behavior, they sat down on Palatine Hill, gazing in different directions. Remus soon saw six vultures, one of the most important and sacred birds in Roman augury. He was shortly outdone by Romulus, however, who saw 12. Clearly, the gods favored Romulus's city-founding plan, because why else would he have seen more vultures than Remus? Faced with the overwhelming logic that 12 vultures are better than 6, Remus was forced to concede defeat.

Palatine Hill it would be. The Eternal City of Rome, named after Romulus himself, was soon built on its slopes. And the ancient tradition of divination from bird behavior became enshrined in the founding of one of the world's greatest cities.

It would not be a happy ending for Remus, who was a sore loser and complained about the placement of the city even after it was so obviously favored by the gods to be built on Palatine Hill. Romulus wasn't having any of that nonsense and promptly had Remus killed.

The vulture doubter became vulture meat.

A Little Bird Told Me

The human condition is one of relentless choice. Decisions face all of us, every day, from the mundane ("Should I take the dog for a walk now?") to the profound ("How should I care for my dying parent?"). It's little wonder that we look for support along the way. After all, decisions have consequences, and you can never know what would have happened if you had chosen a different path. So, for all the long millennia that human beings have endured life on this planet, we've looked for signs to help us navigate this complexity.

Sometimes, as we've searched for answers and omens, we find our gaze drawn upward to the sky. And what do we find there? Birds. This practice of looking for omens in bird behavior, called ornithomancy, has ancient roots in the Mediterranean and was practiced across many cultures, going back at least as far as the Hittites in the thirteenth or fourteenth century BCE. While many cultures practiced ornithomancy, the Greeks and Romans enshrined it in their religious traditions, introducing the concept of "augury," where specific priests were trained to observe and interpret bird behavior. The Greeks likely picked ornithomancy up from the Hittites, but according to the ancient Greek tragedian Aeschylus, the practice of augury was introduced to humans by Prometheus. If his name sounds familiar, it's because he was also the god who gave us fire, angering Zeus enough to get himself bound in an eternal punishment cycle, chained to a mountain with an eagle eating his liver over and over again, day after day. (And yes, it was deeply ironic that the god who gave us auguries also had to endure his liver being consumed by an eagle.)

In *The Birds*, a play by the ancient Greek playwright Aristophanes, a rather creepy chorus of birds explains to human visitors all of the prophetic benefits they have on offer.

Dancing with the birds

Just another day in the life of Prometheus

> *We are your Ammon, Delphi, Dodona, your Phoebus Apollo. Before undertaking anything, whether a business transaction, a marriage, or the purchase of food, you consult the birds by reading the omens, and you give this name of omen to all signs that tell of the future. With you a word is an omen, you call a sneeze an omen, a meeting an omen, an unknown sound an omen, a slave or an ass an omen. Is it not clear that we are a prophetic Apollo to you?*
>
> —Aristophanes, *Birds* 716–22

Birds crop up throughout ancient Greek literature, including in Homer's *The Iliad* and *The Odyssey*, where special priests interpret the flights of birds to guide military strategy. Important characters also offer prayers to the gods for a sign indicating the rightness or wrongness of a decision. Sometimes the gods respond by sending a bird, such as in *The Iliad* when Priam, the king of Troy, prays to Zeus for a sign about his upcoming mission to plead for peace with Achilles:

> *Straightway [Zeus] sent down the most lordly of birds, an eagle, the dark one, the marauder, called as well the black eagle. And as big as is the build of the door to a towering chamber in the house of a rich man, strongly fitted with bars, of such size was the spread of his wings on either side. He swept through the city appearing on the right hand, and the people looking upon him were uplifted and the hearts made glad in the breasts of all of them.*
>
> —Homer, *Iliad* XXIV 315–21

The "appearing on the right" bit is important because birds flying on your right were generally thought to be good signs while those on your left were generally considered bad signs; the left side being associated with unluckiness in Greek philosophy. The baffling arbitrariness of the direction you happened to be facing at the time and the simple fact that shifting your feet slightly in the other direction could change whether the bird was on your right or left was seemingly not accounted for.

The ancient Greeks were deeply invested in ornithomancy, but no one quite perfected it like the ancient Romans. While the Greeks viewed birds as messengers from the gods, the Romans thought that the will of Jupiter, the head of the gods himself, could be interpreted through bird behavior. The Romans applied the same zealous cunning and dedication they used in their military strategy to their interpretation of birds. No state decision was made, no military campaign launched, and no promotion offered without first consulting the birds.

Of course, with so much riding on the outcome of bird interpretation, some sort of skilled intermediary was required. A whole priestly class of individuals arose, called "augurs," whose sole job was to interpret birdsong and bird behavior for signs of divine will. The practice of interpreting these signs—"taking the auspices"—became such an enormously important function of ancient Roman statecraft that we still see echoes of it in our language today: Something is either "auspicious," meaning conducive to success, or "inauspicious," meaning unpromising. To this day, we anoint our highest political office in the United States with an "inauguration."

In fact, you could make a convincing case that the Roman Republic was run by birds.

When any public decision was required in Rome, magistrates consulted auguries before offering declarations of auspiciousness. An official blessing from the gods, arrived at through an auspicious consultation with an augury, was needed before moving forward with, well, pretty much anything. The state was essentially beholden to auspicious bird signs. What if instead of political impasse holding up the budget reconciliation process in the US Congress, it was delayed because no one had yet seen a bird flying in the right direction at the right time?

Augurs carried around something called a "lituus," or a diviner's staff, which was a rod with a spiral at one end that could be used to point toward the part of the sky where they were observing bird behavior. This was as

Augury holding a lituus . . . and carefully watching a chicken

important because the part of the sky from which the bird first appeared, its position relative to the observer, and the direction of its flight were all factors in determining whether something was auspicious or not.

These priests had elaborate rules and protocol to arrive at their avian interpretations. There were also careful record-keepers, filling books with notes on past interpretations as well as details of the rituals and prayers needed to prepare oneself for correct auspicious interpretation.

One rather large challenge to auguries: Wild birds are not reliable. What happens when you need a sign from an eagle and none are flying around that day? Eventually, the Romans were forced to turn to a bird that could be more reliably consulted. This particular bird had several advantages not seen in the wild: It was slow, easily controlled, and seemed rather... unintelligent.

Without further ado, we present to you the humble chicken, great prophet of Ancient Rome. The chicken's comparative lack of intelligence was viewed as a major bonus for omen-catching because the logic, such as it was, was that an empty mind was more receptive to godly influences. (Though, as scientists would later discover, the chicken is a lot more intelligent than the Ancient Romans gave them credit for.) Indeed, the more empty-headed the bird the better, because it was less likely to bypass the will of the gods in pursuit of its own ends. Which was how the humble chicken became just about the perfect bird to receive omens.

Flocks of these familiar barnyard fowl were attended to by an official Roman sacred chicken keeper, dubbed a "pullarius." (To be a pullarius in Ancient Rome meant spending all your days herding, feeding, and, yes, interpreting chickens.) Wherever the Roman army went, a pullarius and a cage of chickens came along for the ride.

Omens from the sacred chickens, raised specifically for this purpose on the Island of Negroponte (present-day Euboea), primarily provided interpretations based on two behaviors:

1. How quickly or slowly they came out of their cages.

2. How much enthusiasm the chickens mustered for the grain and water that were presented to them by the pullarius.

If the chickens emerged too slowly from their cages, it was a bad sign. Similarly, if they showed little interest in their food or, worse still, refused to eat at all, it was a very bad sign indeed. If, however, they came out quickly it was a good sign. And if, on top of that, they began to greedily eat, stamping their feet and scattering their food about, it was a very good sign.

Of course, it wasn't lost on the Roman pullarii that it was fairly easy to manipulate the chickens into an auspicious reading. Chickens like to eat. If you deny them food for a while and then suddenly provide them with a bunch of it, they'll gobble it up. So skipping a feeding or two ahead of a reading a pullarius wanted (or was paid) to make auspicious could all but guarantee a positive outcome.

The great and noble chicken, predictor of all things

Even so, omens from the sacred chickens were not to be trifled with. They were meant to be taken with deadly seriousness. Cicero offers a cautionary tale of the Roman general Publius Claudius, who made the mistake of disregarding the omens from the sacred chickens. As relayed in *De natura deorum* (*On the Nature of the Gods*):

> *Shall we remain unimpressed by the tale of the presumptuous conduct of Publius Claudius in the first Punic war, who, when the sacred chickens, on being let out of the coop, refused to feed, ordered them to be plunged into the water, that they might, as he said, drink, since they would not eat? He only ridiculed the gods in jest, but the mockery cost him many a tear (for his fleet was utterly routed), and brought a great disaster upon the Roman people.*

Cicero was recounting the story of the Battle of Drepana (249 BCE), fought between Rome and Carthage off the coast of Sicily in the First Punic War. Publius Claudius was the commander of the Roman naval fleet and had blockaded the Carthaginian holding of Lilybaeum (present-day Marsala). He had the Carthaginian fleet pinned down and wanted to attack. However, as

Parrot Predictions

For centuries, parrot fortune tellers in south India would employ rose-ringed and Alexandrine parakeets to help tell fortunes. The practice, which is dying out, was particularly prominent in Tamil Nadu and Andhra Pradesh, as well as in Pakistan and Singapore. It works like this: A fortune teller sits beside a well-traveled road with a sign advertising parrot fortunes. If someone engages his services, the client sits in front of the fortune teller and awaits their fate. The fortune teller then sets out a pack of 27 Tarot-like cards, each with an emblem of a Hindu deity. He then releases a trained parakeet from a cage, and it walks over to the cards and selects one, picking it up in its beak and delivering it to the fortune teller. The fortune teller then interprets a fortune for the client based on the auspiciously selected card.

Declining patronage and increased enforcement of conservation and environmental laws designed to protect species like the parakeets have led to a steady decline in parrot fortune-telling. The field, however, is not without its celebrities. Mani the parakeet, an assistant to a parrot fortune teller in Singapore, rose to international fame in 2010 when he correctly predicted the outcomes of all the World Cup quarter- and semifinals. Mani, however, incorrectly guessed the final game of the cup. (Incidentally, the final game of the cup was correctly predicted by Mani's primary animal oracle rival that year: Paul the Octopus.) You can't win them all.

Parrot fortune-telling in action

Publius Claudius about to make a big mistake

was the case for all Roman military and civic decisions in that era, he needed to first receive an auspicious sign from the gods. You can imagine his disappointment when the chickens seemed to think it was a bad idea to attack the Carthaginians. Publius Claudius had a sure thing on his hands: The Carthaginians were trapped between his fleet and the shore. So he ignored the omens, drowned the chickens, and then launched a surprise attack by night.

But Publius Claudius's fleet became scattered in the dark and the Carthaginians were able to spring the ambush. Then they turned around and exploited the same advantage just lost by the Romans. The Carthaginians now had the Romans trapped between their fleet and the shore. And they summarily crushed them.

It was the greatest Roman naval defeat of the entire war, with about 90 percent of its fleet sunk or captured and 23,000 men killed or taken prisoner.

Publius Claudius may have drowned the sacred chickens, but the chickens got their vengeance from beyond their watery grave.

Birds of a Feather

The fall of the Roman Empire in 476 CE also led to the fall of chicken consultation in statecraft. (Though, to be fair to the Romans, by the time Cicero and his contemporaries arrived on the scene in the late first century BCE, chicken consultation had largely been delegated to a performative function to appease less-enlightened members of the late Roman Republic.) While the progress of Western civilization was more or less halted for the next 1,000 years, there was at least one advancement in the ensuing Dark Ages: Chickens went back to being chickens again.

Instead, agricultural cultures in Europe (i.e., everyone for the next 1,000 years) began observing bird behavior and migration patterns to help anticipate weather or seasonal changes and determine appropriate times to plant or harvest. Birds came to be viewed as weather prophets, rather than messengers from the gods.

> Sailors observed of seabirds,
> *When men-of-war hawks fly high, 'tis sign of a clear sky;*
> *When they fly low, prepare for a blow.*
>
> Farmers observed of swallows,
> *When the swallow buildeth low*
> *You can safely reap and sow.*

While not all of these observations were accurate (there is no evidence that a heron flying down a creek reliably predicted there would be no rain that day, or that when woodpeckers pecked low on tree trunks, a hard winter was to be expected), some actually had scientific backing. Sayings also developed around the early arrival of migratory birds from the far north. When such birds flew southward without lingering very long in their usual stopping places, people anticipated a harsh winter. And they were often right. When birds flee early from excessive cold or snow in the far north, it typically indicates a harsh winter has set in and is likely in store as well for more southerly locations.

There are several science-backed predictions we can make by watching birds today:

1. **Low-flying birds can indicate an approaching storm.** As the air pressure drops with the arrival of a storm system, it's more difficult for birds to fly at higher elevations. In response, birds will descend to lower altitudes, where the air is cooler and denser and thus easier for flying. Flying insects respond the same way, which, while not typically noticeable for human observers, is certainly picked up on by the birds, who find more to eat at the lower altitudes when the air pressure drops.

2. **High-flying birds can indicate fair weather, making the converse of point 1 also true.** As the air pressure rises again in the aftermath of a storm system, birds will start flying higher and higher in the sky.

3. **When navigating at sea, an increase in bird counts can indicate approaching land.** Tristan Gooley, author of *The Lost Art of Reading Nature's Signs* and an expert in natural navigation, investigated this phenomenon and reported that if navigators count 10 or more birds in any random five-minute period, there is a 95 percent probability of land within 40 nautical miles. The closer you are to land, the more within range you are of any non-pelagic bird species, which do not spend much of their life on open water, and thus stay closer to shore.

4. **When navigating at sea, a decrease in bird counts can indicate that you're moving farther away from land.** This is of course the converse of point 3, also demonstrated by Tristan Gooley. If three or fewer birds are counted within a random five-minute period, there is a 90 percent likelihood that land is over 50 nautical miles away. Beyond 50 nautical miles and you are increasingly in the exclusive domain of pelagic birds, or seabirds that spend much of their life on open water.

5. **Birds can predict the arrival of severe weather, often long before humans.** Biologists studying golden-winged warblers in the Cumberland Mountains of East Tennessee were surprised to observe that the birds disappeared entirely just two short days after arriving at their breeding grounds in the mountains. They flew 932 miles away to the Gulf Coast, waited a few days, then returned to the mountains again. Why? The birds took off when they sensed the arrival of a powerful storm system, which, when they left, was still between 250 and 500 miles away. The storm system would spawn 84 tornadoes and kill 35 people. The warblers departed the mountains days before weather forecasters issued their first warnings about the storms.

6. **Bird-nesting behavior can predict the severity of a hurricane season.** Ornithologist Christopher Heckscher accurately predicted the severity of the 2018 hurricane season by watching the nesting behavior of the veery, a small songbird in his native Delaware. Careful analysis conducted over the course of decades revealed that veeries had longer breeding seasons during years with mild hurricane seasons and cut their nesting short in anticipation of harsher seasons. Calling it early on the breeding season and starting on migration gives the birds some flexibility to wait out strong storms before attempting to cross open water to reach their wintering grounds in Brazil. The veeries, it turned out, were better at predicting the severity of the hurricane season than the National Weather Service.

The birds all around us are incredible creatures, acutely attuned to the environment and with the ability to respond to elements in the natural world that are not easily observed by humans. Further study is needed to fully understand the signs birds are able to interpret before we do. One thing, however, is certain: Chickens are far better at egg production than statecraft. So let's keep 'em out of politics.

Polygraph

n the 2017 remake of *Wonder Woman*, American soldier Steve Trevor is bound by a golden lasso, kneeling before Wonder Woman and the women of the hidden island of Themyscira, where he crash-landed.

After unwillingly admitting his identity, he looks at the glowing rope around his body and asks in surprise, "What in the hell is this thing?"

"The Lasso of Hestia compels you to reveal the truth," Wonder Woman tells him.

"But it's really hot!" he complains.

"What is your mission?" demands her mother, Queen Hippolyta.

After a brief moment of resistance, Steve speaks, almost in pain.

"I am a . . ." He pauses, grunting in discomfort, before the words blurt out in a near shriek, like a cork pried loose from a bottle of hot champagne: *"I AM A SPY!"* He shrugs, no longer resisting the lasso. The cork is out. "I'm a spy," he repeats again and again.

If only Wonder Woman really existed. If only humanity possessed that miraculous golden lasso to compel liars to tell the truth. It would utterly change criminal justice (and politics), wouldn't it?

But we do not live in a comic book universe, so we mortals must make do with human-made lie detection instead. Which is perhaps why, in one of the most random connections in history, the creator of Wonder Woman—psychologist William Moulton Marston—also created the prototype of today's polygraph machine.

False: The polygraph is commonly used by fathers on their future sons-in-law.
True: They just wish they could.

Also known as the lie detector test. Everyone has heard of it. It appears in movies, like when Robert De Niro's character grilled his future son-in-law in *Meet the Parents* or when Sharon Stone's character was given the test in *Basic Instinct*. It's used by the FBI and several other governmental agencies to ensure that employees in the most delicate of national security positions are honest characters. People who want the public to believe they are speaking the truth will employ them to prove their honesty, as Dr. Christine Blasey Ford did during her widely publicized allegations that US Supreme Court Justice Brett Kavanaugh sexually assaulted her in the 1980s.

The polygraph carries a lot of weight in the US. A particularly American phenomenon used scarcely outside the US, the polygraph has become part of our culture. Unfortunately, it's generally considered an unreliable test by the scientific community—one that can be fooled with a few learned tricks.

Lying: A Brief History

The history of lie detection is old. In medieval times, the Welsh used what was called a "compurgation" (yep, that's a real word) to prove one's innocence. If accused of murder, for example, the accused could be acquitted via compurgation by gathering people who would testify that the accused couldn't possibly have done such a thing. The required number of witnesses depended on the crime, but in cases of murder, 600 were needed. Which is ridiculous, because what ordinary person knows 600 people well enough to ask them to testify? Particularly back in medieval times?

In the Middle Ages, honest people were thought to withstand torture far better than lying ones, so using a so-called "trial by ordeal" at the time might involve a person being required to lick a burning-hot iron. If their tongue came away unscalded, they were truthful. In the fifteenth, sixteen, and seventeenth centuries, people accused of being witches were subjected to various types of torture to prove their honesty and innocence. Tying someone up and throwing them into the water to see if they floated (a sure sign of sorcery) was called "swimming a witch." The innocent sank—a pretty lethal way to prove one's truthfulness.

Men having a blast as they try to drown a so-called witch. Wales, circa 1876.

Making someone gather a gazillion witnesses of good character or torturing them was obviously far from a perfect means of extracting the truth. In the nineteenth and early twentieth centuries, people tried to ascertain if blood pressure, respirations, and pulse measurements could detect "tension" or emotion. But we still hadn't developed a good test for lying as we entered the twentieth century. However, while studying blood pressure in the early 1910s, William Moulton Marston took note that his wife felt like her blood pressure rose when she was overtly emotional. He then realized that blood pressure changes might correlate with lying.

The theory, which still remains the reason why polygraphs are believed to work, is that when a subject is telling a falsehood about something pretty important, such as stealing a million dollars or killing their spouse, the body expends more energy to lie. In short, Marston came to believe that our bodies simply cannot suppress the stress of deception, so they react involuntarily with elevated blood pressure, or fast breathing, or an elevated pulse rate. Marston became an outspoken proponent for the intermittent blood pressure measuring device he created in 1915, authoring multiple journal articles about it and trying to get it into mainstream criminal work in the court system. He even got himself in an advertisement regarding the veritable superiority of Gillette razors. A random factoid, but we guess that truth in advertising was as important back then as now.

Eventually, Marston's device evolved into what became the modern polygraph, for which John Larson can claim credit because he improved on Marston's device by measuring continual blood pressures. By the 1930s, the machine also measured breathing and "skin galvanic resistance," aka sweating. Except for being slightly more computerized and including the addition of respiration measurements, the polygraph test hasn't changed much since its conception. This feels a little surprising, given that it's performed 2.5 million times a year (as of 2018) and drives a $2-billion-a-year industry.

Polygraphing 101

Here's how it works: Those taking a polygraph are subjected to the comparison question technique (CQT), the most frequently used questioning technique in the world. Introduced by John Reid in 1947, the progression goes like this: First, the tester will talk to you. They tell you how the machine works and what to expect, and then they ask if you have medical conditions or take medicines. But they are also getting to know you, studying you as you acquaint yourself with the process, and letting you know that the test is incredibly accurate—so you'd better comply. The intimidation factor is no small thing.

Two stretchy cords are then placed around your chest, above and below your heart, to measure breathing rates and sometimes bodily movements. A blood pressure cuff on the arm measures pulse and pressures. And two finger or hand pads will measure sweating, or electrodermal activity. At this point, the examiner will ask a bunch of questions, such as verifying your address and name, but also others they expect you to lie about. Things like "Have you participated in a sex act you're ashamed of?" or "Have you ever stolen something at a job?" These are things that everyone has probably

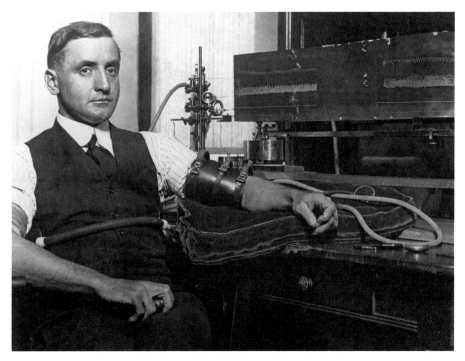

Murder suspect Henry Wilkens is cleared of murdering his wife by a machine that is not scientifically accurate.

done but will feel uncomfortable admitting. These are part of the "acquaintance" or "stimulation" tests, so the examiner can immediately begin to see the physiologic differences between when you lie and when you tell the truth, or if you use "countermeasures," or methods to conceal your lying. They are the control questions.

The examiner will then mix in questions about the issue they are interested in—if, for example, you killed your neighbor because you didn't like their sloppy lawn maintenance, or if you plan on selling national security secrets to Russia. At the end of the test, the examiner will pore over the testing results and draw their conclusions. The whole thing takes about an hour.

If the examiner is lucky, the interviewee will be so upset by lying in response to the questions that they'll simply confess to whatever crimes or misdeeds they've committed. When this happens, examiners have proof that their test has 100 percent accuracy—except when an innocent person confesses to a crime, which certainly occurs.

In controlled environments, such as research with volunteers, accuracy can be measured but still vary widely, from 50 to 90 percent. Controlled lab settings are definitely not the same thing as questioning someone whose entire life is on the line if they murdered someone. So there are many who believe that these controlled environments can't be included in research on polygraph accuracy.

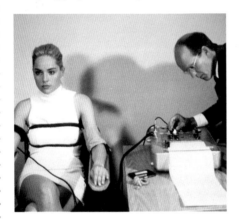

Catherine Tramell demonstrates how to ace a lie detector test in Basic Instinct.

But if the test itself has significant life consequences, there's also a very good chance that the person being examined might be nervous when asked a troubling question. Perhaps you didn't kill your friend, but being accused sure does make your blood pressure shoot up. Polygraph examiners claim they have ways to tell the difference. Nevertheless, when you get heart palpitations because you're asked if you've committed acts of bestiality and you have not, this can result in a false positive test and a lost job opportunity if you wanted to work for the CIA.

Furthermore, there are countermeasures people use to fool polygraph tests, which raises real questions about testing accuracy. You can train yourself to beat the polygraph. In fact, a National Security Agency whistleblower, Russell Tice, stated that he took somewhere between 12 and 15 polygraph tests during his 20-year career there. He would bite his tongue during the control questions in the stimulation test phase, so that if he lied later on, his physiological measurements would be calm in comparison. Or he would daydream to calm his nerves.

In some ways, the polygraph isn't so much a test for falsehoods as a test for fear. Ideally, it measures physical signs of discomfort, or the fear of being caught in a lie. So it's not a true lie detector test. Professional polygraph examiners naturally dislike this distinction. The machine is simply a tool, but it's the human behind it who makes the assessment or opinion. And humans are subjected to many types of biases, including those around gender, race, sexuality, socioeconomic status, and more. In 1987 during a Senate subcommittee hearing about polygraphs, the New York attorney general testified that "The [polygraph] operator's prejudices, moods and feelings can strongly influence and even determine the outcome of the test. For example, we have received complaints about a polygraph operator who consistently fails a much higher percentage of black subjects than white subjects." Reports abound of law enforcement agencies that fail more black applicants than white ones, though the data is hard to mine because most agencies don't publicly post their information on the subject.

In the absence of a purely voluntary confession, there remains no foolproof way to check the accuracy of a polygraph test. Many of the so-called cues, like voice pitch changes, shifting in a chair, crossing arms, or fidgeting are not clinical signs of lying. Discomfort, perhaps, or nervousness, but not a reliable indicator of actual mendacity.

The Truth About Polygraphs

Many polygraph experts claim accuracy rates of 90 percent or above. But even if those numbers were correct (and studies that cite this percentage are routinely debunked as methodologically inadequate), think of the consequences. Assuming a robust 90 percent accuracy rate, that means 10 percent of the time the polygraph test is wrong. In large-scale screenings, such as for employment, this can translate to a large number of false positives—meaning people being accused of lying when they are not. For mass screenings, this usually results in a potential job in law enforcement or at a government agency being lost—and those test results could then be shared with other agencies. Taken a step further, what happens when that applicant is asked, "Have you ever failed a polygraph test?" on their next application and must answer "yes"? It doesn't matter if you were telling the truth the whole time, because the polygraph stain is unwashable. In criminal justice scenarios it's even worse, because an innocent person could be accused of lying, or a liar could possibly get away with murder.

Gary Ridgway, mass murderer and lie detector liar

One of the most infamous examples of this is the so-called Green River Killer, Gary Ridgway. A prolific and horrific serial killer, Ridgway was convicted of 49 murders, eventually confessing to the killings of 71 women. Some estimate the number is closer to 90. In 1984, the police assembled a profile fitting Ridgway and brought him in for a polygraph test. He passed. His DNA was gathered years later, and he was eventually arrested for the murders in 2001—a full 17 years after he passed his polygraph test, during which time he was confirmed to have murdered six women and was suspected of killing an additional six more.

Several spies have been known to beat the polygraph test, much to the chagrin of the government that was unknowingly harboring them. Leandro Aragoncillo was a spy within the White House during the Bill Clinton and George W. Bush administrations. Karl Koecher, a Czechoslovakian spy for Russia who infiltrated the CIA after passing the entrance polygraph exam, was another, as was Ana Montes, who spied for the Cuban government while working in the Defense Intelligence Agency and passed polygraph tests.

Author Ronald Kessler, right, interviews Karl and Hana Koecher. Karl was a Czech spy, triple agent, and lie detector beater.

What's probably most telling about the reliability of the test is that two-thirds of psychologists polled in the American Psychological Association believe that CQT tests are not supported by scientific evidence, and three-quarters do not think CQT tests should be admissible in court. The US Supreme Court agreed in 1998 (*United States v. Scheffer*), deciding that polygraph tests cannot be used at the federal court level because of their unreliability (with rare exceptions).

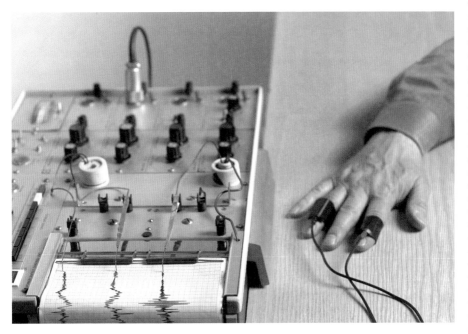
Finger monitors on a lie detector test measuring sweating

On the state level, there is a troubling lack of consensus on whether polygraphs are admissible evidence in the court system. For example, New Mexico allows them only if both parties agree; Massachusetts bars them completely. Right now, about half of US states allow the polygraph to be used as court evidence, though the debate over the test's scientific credibility can muddy its evidentiary usage on a case-by-case basis. Law enforcement continues to use it for employee screening, but in criminal investigations, they cannot force a person to take one because it would be violating the Fifth Amendment that protects against self-incrimination.

In 1988, the federal Employee Polygraph Protection Act prohibited the use of polygraphs in private employment because of their scientifically proven weaknesses. That didn't prevent the government from using them, however. About 12,000 tests are performed annually by the FBI, DEA, and Bureau of Alcohol, Tobacco, Firearms and Explosives in screening prospective employees and in criminal investigations.

And there is of course the economics to consider. The modern polygraph machine isn't cheap—each costs several thousands of dollars, and each test costs approximately $700 to administer. Tuition for the required 400-hour training can run $4,995. Which is a lot of hours and money for competency in what is a very subjective test. Twenty states do not even require a license to perform a polygraph test, which calls into question the standards for the testers themselves in those places.

From 2002 to 2005, the FBI budgeted $24.9 million to support its polygraph program—a decent chunk of taxpayers' money. This was done despite a 2003 National Academy of Sciences (NAS) report on the scientific validity of polygraph testing, which concluded that the research to support CQT was faulty and of low quality. Though the CQT's accuracy was definitely better than flipping a coin, there was no standard error rate. And anyway, "slightly better than chance" as an accuracy doesn't sound like something you'd want to bet your life on, or rely on when it comes to your career or keeping someone inside (or out of) prison for a crime.

Conversely, many within the polygraph community think too much has rested on the shoulders of the NAS's decision back in 2003 about the unreliability of the polygraph. After all, it's been about two decades since that report. So the American Psychological Association published an updated one by William Iacono and Gershon Ben-Shakhar that examined the current status of forensic lie detection using the CQT in 2018. The report closely examined what new data had accumulated since 2003, including more controlled laboratory type studies. Their conclusion?

There was no high-quality research on polygraph accuracy since the 2003 report, and its original conclusions should stand. Translation: The polygraph profession's claims that its testing is highly accurate are still unfounded.

Back in 1915, William Moulton Marston had his heart in the right place in hoping for a machine that would shine the light of truth in the darkest, most secret corners of the world (and our minds and bodies). Unfortunately, the modern polygraph is about as fictional as Wonder Woman's golden lasso.

Superstitions Hall of Fame

The Psychology of Superstitions

If there is a death in the family, you should inform honeybees on your property. Otherwise, they will die.

For weddings, wear "Something old, something new, something borrowed, something blue." Or this one: "Marry in black, you'll wish yourself back."

Two mirrors facing each other and reflecting into each other are the "doorway to the devil."

Superstitions exist everywhere around the world. It's quite likely you believe in at least one or two yourself. But why do we put faith in such arbitrary, irrational things?

The answer is that somewhere, someone found a simple cause and effect to superstitions and then decided they ought to be unbreakable. Did your favorite team win a game after you wore a special T-shirt? Better wear it every time! Because why tempt fate? And superstitions are often harmless. Avoid stepping on a crack—why risk a vertebral fracture of your mom's already aching back? And finally, confirmation bias. If you've never suffered a terrible accident by avoiding walking under a ladder, then clearly that's a good thing.

But another thing superstitions give us is power. Power over a world filled with fear, unpredictability, and uncertainty. Superstitions are habits that have become part of our lives. They're hard to ignore, and they're embedded in each and every one of our cultures. And let's be honest: They're kind of entertaining—and usually cheap and harmless (unless that team shirt starts to really stink, because you decided washing it would rob it of its luck). So until our rational side takes 100 percent control of our lives, we'll keep saying "Jinx!" every time we say the same thing simultaneously.

Knocking on Wood

Who hasn't knocked on wood to stop the universe from delivering the bad thing you just spoke about? Even if there's no wood around, you could touch a piece of paper. It was wood once, anyway, wasn't it?

Knocking on wood, or even saying "touch wood," are both methods to avoid a bad fate after speaking about something, positive or negative. An example would be, "I should be fine flying directly into this tornado, right?"

knock, knock, knock

The origins may lie in an old game of tag from 1800s England called "Tiggy, Tiggy Touchwood" in which a player, "Tiggy," chases others, who are safe if they are touching a branch or a tree trunk. "Tig" is a variant name for the game "tag."

A common origin hearkens to Celtic peoples who invoked the spirits of trees. Similar superstitions exist across the globe. In Egypt, مسك الخشاب or "emsek el-khashab" means "hold the wood" and is said as a way of acknowledging good things, hoping they continue, and preventing envy. In Italy, a version is "tocca ferro," or touch iron, spoken when anything related to death is seen.

Does it actually help? Invoking humbleness and good luck can't hurt anyone. Also, it's been scientifically established that childhood play like tag is definitely a good thing. Touch wood.

Tiggy, Tiggy Touchwood!

Black Cats

Whether or not you're a cat lover, no one can deny that a sleek black cat with almost preternaturally glowing yellow eyes is a striking visual. But it's also commonly associated with bad luck and the paranormal. Just think of Halloween, which is rife with black cats and witches. In Western tradition, the black cat has been cited as the witch's "familiar," or a guardian or protector. Witches were thought to be able to shape-shift into black cats as well. The fear of having a black cat cross your path arose in several areas. In Germany, it's bad luck if the cat walks from right to left—but good luck if it walks from left to right. Similarly, eighteenth-century pirates believed that if a black cat walked onto your ship and then left the ship, it meant that ship was doomed to sink. In Great Britain, a black cat walking away was also "taking the good luck away" from you.

Interestingly, many cultures think of black cats in a very positive light. In ancient Egypt, they were symbols of the divine. In Japanese folklore, black cats are considered good luck. Scottish lore says that a black cat entering your home means coming prosperity.

These days, most people don't believe that cats portend evil things. But the spooky aura

You shall not pass!

around them doesn't help them get adopted from shelters. The UK RSPCA has deemed September 27, near Halloween, as National Black Cat Day, to spread awareness that black cats take twice as long to rehome. Some people think they're less adoptable because they're less photogenic. In that case, you can blame photo-happy social media and not superstition for that bad luck.

Wishbone

Around Thanksgiving, many Americans practice the tradition of playing tug-of-war on a single poultry bone to make their dreams come true. That little bone in the turkey or duck or goose, nibbled free of meat, is known commonly as a wishbone but is technically referred to as the furcula. Sounds like a cuss word, doesn't it?

"Furcula!"

But we digress. Furcula means "little fork" in Latin, which also sounds like a cuss. Most birds have this, and it supports their rib cage and body while they fly. The human version of the furcula would be a pair of clavicles fused together in the middle.

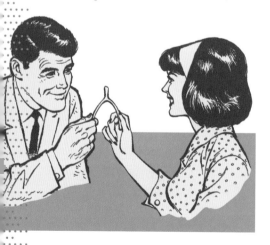

Making that furcula wish come true

Around the late medieval period, the appearance of these wishbones was used to divine the coming winter, just as Groundhog Day is used today to help divine the coming spring. But the tradition of two people pulling on the wishbone began in the early seventeenth century. The person with the larger portion after the break is gifted with a wish come true. Interestingly, the wishbone back then was called a "merrythought." Too bad our clavicles aren't called "jubilant cogitations." That would look strange on an X-ray radiologist report.

Lucky Horseshoes

The concept of lucky horseshoes is well-known to many. At the very least, you might have eaten them in the form of crunchy marshmallows in your Lucky Charms breakfast cereal. One origin of the lucky horseshoe might have come from St. Dunstan, a second-century English bishop and the patron saint of blacksmiths. Legend has it that the devil visited Dunstan, asking for horseshoes for his horse. Instead, Dunstan tricked the devil by shoeing Satan's own cloven hooves with burning hot shoes. The devil begged him to remove them, and Dunstan did, but not before he made the devil agree not to visit homes where a horseshoe was hung.

Satan's shoe?

Some think that horseshoes must be hung pointing upward, to "catch" the good luck around. Others think they should point downward, to spread the good luck to passersby. And still others believe that horseshoes should be nailed in with seven nails, as that number imparts luck as well. Finally, there is also the fact that horseshoes were traditionally made with iron, which in fairy lore is a deterrent against evil fae creatures. The lore of lucky horseshoes has managed to persist over centuries, though most probably have no idea that it has to do with the devil wearing hot shoes (and not the Prada kind).

Unlucky Friday the . . . 17th?

We're all familiar with the unlucky number 13, but did you know that in Italy 17 is the bad luck number? And there are some good reasons why.

First of all, the roman numerals for 17 are XVII, which is an anagram for the Latin word "VIXI," meaning "I have lived." In other words, "I am now very dead," an inauspicious association. Also, the biblical great flood occurred on the 17th day of the month.

As for Fridays? Well, Jesus died on Good Friday, for one thing. There's also an Italian saying that goes, "Né di Venere, né di Marte non si sposa, né si parte, né si dà principio all'arte." Or "Neither from Venus nor from Mars [Friday nor on Tuesday] does one marry, leave, or start something."

As a result of these superstitions, some Italian airlines don't have a seventeenth seat. A car model R17 by Renault is sold in Italy as the R177, so as not to tank sales. (International marketing research is important! After all, Chevy was not so smart when it tried to sell the Nova car model in French- and Spanish-speaking countries, where the car's name literally means "Doesn't go.") Also, people might carry charms or hang them in their households to shield against those inauspicious Fridays and 17th days of the month.

Given that some of this fear comes from the Latin anagram for VIXI, remember that most anagrams aren't nearly so scary, even when those have to do with numbers. Take "eleven plus two," which is an anagram for "twelve plus one." Nifty, somewhat random, but not scary at all.

The Number 13 and Numerology

Number superstitions exist all over the world (see directly above). In China, Vietnam, Korea, and Japan, the number four is quite inauspicious. This is because the word for the four, sì (四), sounds similar to sǐ, or "to die" (死), in Cantonese. The Book of Revelation in the Bible references the terrifying four horsemen of the apocalypse, thought to be the personified figures of War, Death, Famine, and Conquest. Many people link the number with good things, like the four-leaf clover, a rare and lucky find. The number 12 has been considered a good number for a variety of reasons (12 months in the calendar year, 12-hour divisions in the day). Seven is a number that so many people also love. Seven days of the week, seven days in which the Bible says the world was created, and the so-called seventh son of a seventh son ought to be super-duper special. In Islam and Judaism, there are seven heavens.

But 13 has always had a particularly dark place in the world of superstition. The 13th person at the Last Supper was Judas Iscariot, who betrayed Jesus. According to Norse legend, trickster Loki was the 13th guest to show up at a dinner in Valhalla, followed by Loki tricking the god Hodr into killing his brother Baldr (god of goodness and light) using a poisonous mistletoe-tipped arrow. Some have associated the 13th member of a witches' coven with the devil. It's such a common American superstition that many buildings in the United States simply don't have a 13th floor. One company, Otis Elevator, noted that 85 percent of its buildings didn't have a 13th floor. And so many people are so scared of the number 13 that there's a word for that: triskaidekaphobia. Fear of Friday the 13th is an even bigger mouthful: paraskevidekatriaphobia.

But for every bad association with 13, there's a good one. In Italy, it's lucky for football game pools. A baker's dozen of anything bought in a bakery can't be anything but good (unless you're gluten intolerant). In Judaism, it's an important coming-of-age year. There are 13 postures in tai chi, which has been clinically proven to help prevent falls in the elderly. Also, it's Taylor Swift's lucky number.

Judas, the thirteenth guest at the Last Supper, betrays Jesus with a kiss.

Ultimately, when it comes to numbers and good or bad luck, it's the flip of a coin depending on what you believe. Or what pop star you idolize.

Crossing Fingers

Chances are, you've done it: crossed your index and middle fingers together to bring some added luck to your life. Or maybe you've hid crossed fingers behind your back when telling a lie. Many of us have also used it in our emoticons to friends on texts or social media, wishing good things.

An effort at neutralizing your lies

So where did the finger cross come from? The true beginnings are obscure. Some believe that the crossed fingers were emblematic of the Christian cross, and people in sixteenth-century England used it as a means of warding off bad luck. The gesture was also relatively unnoticeable, which might have been helpful in not attracting dangerous attention from anti-Christian populations. However, many find this origin story inaccurate or unprovable. The fingers don't actually look like a crucifix, and historical references are lacking. Early-twentieth-century references are plentiful, however. Similarly, the origin of using the gesture to absolve yourself of a little lie is equally obscure.

In other countries such as Vietnam, the crossed fingers are considered to look a little bit like genitalia and are considered a rude gesture (akin to flipping the bird).

Does it help? Does it hurt? That's up to you. But sometimes, these superstitions take on a cultural significance of their own. Just as an after-sneeze "bless you" doesn't bequeath any true health benefits but is considered polite behavior, a few crossed fingers emoticons can deliver a message of goodwill that can be appreciated from right next door or thousands of miles away.

Just don't send them to your friends in Vietnam.

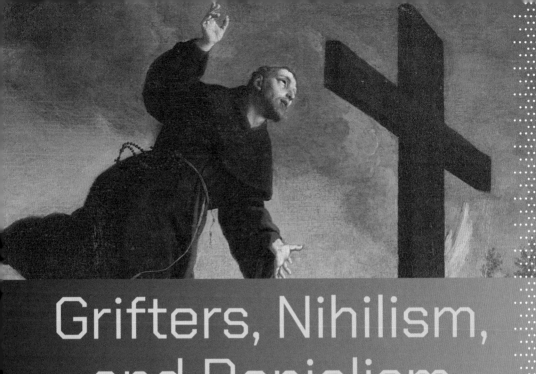

Grifters, Nihilism, and Denialism

Levitation

H**umans can't fly.** It's simply not in our DNA. The last common ancestor we shared with birds was a tetrapod vertebrate, which lived 330 million years ago. It's been another 80 million years since we had a common ancestor with the only flying mammal: the bat. So we have many millions of years of evolutionary history separating us from the flying creatures of the world. During those critical millennia, birds and bats developed the aerodynamic wonder machines we call wings, which are (as you may have observed) a critical component for the gravity-defying act otherwise known as flying. Meanwhile, we humans were hard at work developing opposable thumbs. Not helpful for flying, but useful for just about everything else. You can't have it all, apparently.

So here we are, Earth-bound and hemmed in by gravity but really good at developing tools. For most of us, this probably seems like a reasonable trade. Sure, we can't leap from a building and fly through the air, but we did invent engines, gain an understanding of aerodynamics, and even build airplanes and spaceships.

Nevertheless, there have been people throughout history who have claimed to be able to levitate or even fly, in flagrant defiance of every law of physics. The gravity of the Earth itself pulls everything—and everyone—toward it. You need a lot of lift to overcome both the force of Earth's gravity (a constant acceleration of 9.8 meters per second squared) and your own weight in order to take to the air. What can supply that much lift? Spiritual ecstasy, some would argue.

Many of the people who have claimed to levitate were united by a common thread: a deep religious faith. Levitations in the midst of spiritual ecstasy have been reported in the historical record as far back as the second century BCE, and for a while the medieval Christian church was practically overrun with saints who couldn't keep their feet on the ground.

House of the Rising Nun (and Flying Friar!)

There were many reported incidents of monks and nuns levitating in the medieval Catholic Church, but the most striking cases of saintly levitation were those of St. Teresa of Ávila and St. Joseph of Cupertino, who (yes) rise above the usual milieu of levitation reports because both of their flying episodes were witnessed by a spate of reliable observers.

Thanks in part to her prolific writings, St. Teresa of Ávila (1515–1582) is remembered as one of the foremost Christian mystic saints. Early in life she joined a Carmelite convent—but soon grew impatient with what she viewed as the lax spiritual vigor of the Carmelites of her day. She then left her convent and began traveling around Spain, founding a reformed Carmelite order with strict, devout rules. Teresa began spending much of her day in prayer, and it was during one of these all-day prayer sessions that a curious thing happened: Teresa began to levitate. Then it happened again. And again.

Surprised and not a little embarrassed, Teresa tried to keep her levitations secret. It didn't work. Soon, she began levitating while attempting to conduct services, relying on her fellow sisters to hold her down to stop her

St. Teresa of Ávila

from flying. She also sometimes levitated on receipt of holy communion. Teresa herself was hesitant to talk about these episodes of religious ecstasy, but at least 10 firsthand accounts survive attesting to her frequent levitations.

"I was in the choir waiting for the bell to ring, when our Holy Mother [Teresa] entered and knelt down for perhaps the half of a quarter of an hour," recalled Sister Anne, a nun in Teresa's order, during Teresa's beatification proceedings after her death. "As I was looking on, she was raised about half a yard from the ground, without her feet touching it. At this I was terrified, and she, for her part, was trembling all over. So I moved to where she was and I put my hands under her feet where I remained weeping for something like half an hour while the ecstasy lasted. Then suddenly she sank down and rested on her feet and turning to me, she asked who I was and if I was there for a long time. I responded, yes. She ordered me under obedience to say nothing of what I had seen."

Teresa eventually overcame her hesitations about discussing her mystical experiences, composing an autobiography, as well as two other books, that became foundational texts in Christian mysticism. The autobiography, still in print today, contains several vivid passages describing her levitation experience.

"When I tried to resist these raptures, it seemed that I was being lifted up by a force beneath my feet so powerful that I know nothing to which I can compare it, for it came with a much greater vehemence than any other spiritual experience and I felt as if I were being ground to powder. It is a terrible struggle, and to continue it against the Lord's will avails very little, for no power can do anything against his."

Teresa prayed that these rapturous levitations would stop, as worry over them was exhausting her. Seemingly her prayers were answered, as

St. Joseph of Cupertino in action, startling fellow monks, angels, and local farmers all at once

the levitations indeed declined significantly later in life. Modern-day medical historians think there may have been something else going on here: temporal-lobe seizures. What Teresa identified as "raptures" may have actually been seizures brought on by undiagnosed and untreated epilepsy. Indeed, upon rereading Sister Anne's account of Teresa's levitation, it sounds a lot like how someone might describe an epileptic seizure if they had never seen one before and had no basis for understanding it.

Teresa was followed in the air by perhaps the most famous religious levitator, the so-called "flying friar" St. Joseph of Cupertino (1603–1663), who could hardly keep his feet on the ground. Joseph was an ascetic, the sort of religious fanatic who wore hair shirts, whipped himself, and put a bitter powder on the small sampling of unflavored vegetables or beans he would eat after fasting for days. He also may have had an undiagnosed mental or neurological disorder; he was nicknamed "the Gaper" for his vacant, slack-jawed expression during his youth. After joining a monastery, Joseph, too, began to experience levitations in moments of religious ecstasy, a state of being that was easily provoked in the monk. Hardly a mass seems to have passed without Joseph levitating at some point during it, or while praying in private preceding or following the service.

Joseph's reputation as the "flying friar" spread across Europe, attracting curious travelers from far and wide who came to see Joseph in action. The monastery had to install temporary housing just to shelter the crowds. More than 70 of Joseph's levitation episodes were recorded by witnesses, including Pope Urban VIII himself.

In reviewing accounts of Joseph's levitations, later investigators concluded that many of the eyewitness reports were subject to gross exaggeration or were written years after he had died in 1663. Investigators also reported that many of the reported levitations originated from a leap, rather than a sudden elevation from a sitting or standing position. Joseph may have actually just been a really good—and really enthusiastic—jumper, surprising those around him with his sudden leaps of religious joy.

Whether jumper or levitator, Joseph was canonized in 1767, becoming St. Joseph of Cupertino, the patron saint of air travelers and pilots.

Spiritual Runnings

Levitation by spiritual ascetics was not the sole purview of Christian religion. Reports of levitating Hindu fakirs in India and Buddhist monks in China were sometimes breathlessly reported by early Western travelers. Louis Jacolliot, a French barrister and judge in colonial India, traveled across the country while researching his book *Occult Science in India and Among the Ancients*. In Benares, he encountered a holy man named Covindasamy who was able to communicate telepathically, perform telekinesis, and, naturally, levitate.

"Leaning upon the cane with one hand, the Fakir rose gradually about two feet from the ground. His legs were crossed beneath him, and he made no change in his position, which was very like that of those bronze statues of Buddha that all tourists bring from the far East, without a suspicion that most of them come originally from English foundries.

"For more than 20 minutes I tried to see how Covindasamy could thus fly in the face and eyes of all the known laws of gravity; it was entirely beyond my comprehension; the stick gave him no visible support, and there was no apparent contact between that and his body, except through his right hand."

A few decades later, while traveling in China, Alexandra David-Neel, a French writer and explorer who had converted to Buddhism, wrote about witnessing lung-gom-pa runners. These were monks who, after completing a rigorous spiritual training program, were able to run extraordinary distances without stopping. Their fleetness of foot gave the impression that they were quite literally flying.

"I noticed, far away in front of us, a moving black spot which my field-glasses showed to be a man," wrote David-Neel in her book *Magic*

Levitating man trick

and Mystery in Tibet. "I felt astonished. Meetings are not frequent in that region [Chang Thang in northern Tibet] . . . But as I continued to observe him through the glasses, I noticed that the man proceeded at an unusual gait and, especially, with an extraordinary swiftness . . . The man did not run. He seemed to lift himself from the ground, proceeding by leaps. It looked as if he had been endowed with the elasticity of a ball and rebounded each time his feet touched the ground."

Lung-gom-pa runners could run for 48 hours without stopping or put in 200 miles in a single day. They were able to achieve their extraordinary endurance and lightness of foot only after years of secluded meditation in hermitages, practicing specific breathing exercises and chanting particular mantras. "Lung" (pronounced "rlun") signifies both "air" and "prana," or "spirit," the vital energy inside of a person. "Gom" (pronounced "sgom") means "meditation." A "lung-gom-pa" was a person who had mastered the art of controlling their energy through meditative breathing, achieving a state of spiritual enlightenment that in turn allowed them to transcend the physical limits of their bodies.

Alexandra David-Neel remains the only person to have recorded an eyewitness encounter with a lung-gom-pa runner. If the training still takes place today, it happens in secret.

Floating Mediums

Daniel Dunglas Home was a famous nineteenth-century medium who boasted of a uniquely wide variety of psychic skills, including speaking with spirits of the dead and levitating in front of witnesses. Combining his potent psychic skill-set with a healthy dose of natural charisma, Home fully flowered in the nineteenth-century Spiritualist movement that swept across Europe and the US. By the time he was in his early 20s, he was already world famous, consorting with

Daniel Dunglas Home, in splendid Highlander attire

kings and queens and other heads of state, as well as prominent leaders in science, religion, politics, and literature.

Everyone wanted to attend a Home séance. And for a good reason: He really knew how to put on a show. In addition to the usual table-rapping and -lifting, which were understood in the context of Spiritualism to be communications from the afterlife, Home would frequently levitate in front of his audience. On a good day, you might even get some miraculous healings; eerie, glowing lights; and demonstrations of superhuman strength. The séance would then close with the ringing of bells; the appearance of "spirit hands," which were glowing, luminescent, disembodied hands; and a few distant notes on the accordion.

Home, once again rising above the fray.

In London during the mid- to late 1800s, you couldn't throw a cat and not hit a medium who could produce some good table-rapping for you, but someone who could actually levitate? That was special.

Home's levitations earned a special place in history because none of his contemporaries could explain them. Numerous breathless eyewitness accounts exist, some from quite prominent attendees of Home's séances, where he levitated in front of them.

One of Home's most infamous levitations happened in 1868 in London, when he not only levitated up from the séance table but then "flew" out of the window of a five-story building and back in again through another window. Three reputable witnesses, Windham Wyndham-Quin, 4th Earl of Dunraven (styled Viscount or Lord Adare); James Lindsay, 26th Earl of

Crawford (Lord Lindsay); and Charles Wynn all testified to having seen this remarkable feat.

Lord Adare in particular had become something of a Home groupie (and, possibly, lover). From 1867 to 1869, he followed Home as he traveled around Europe, supporting the medium financially and bearing witness to numerous levitations. In addition to the famous "out one window, in another" levitation, Adare also saw a rare outdoors levitation. (Outside levitations were given extra weight in accounts of the phenomena because of the decreased ability to disguise some sort of trickery.)

"Presently we saw him approaching and eventually raise off the ground . . . he floated by in front of us, at a height which carried him over the broken wall, which was about two feet high. There could not be a better test of being off the ground, for as he crossed the wall, his form was not in the least raised, but the movement was quite horizontal and uniform. The distance that we saw him thus carried must have been at least ten or twelve yards."

Home's levitations generated significant controversy—and significant supporters. Several scientists who were allowed access to Home during his séances came away convinced they were witnessing a real phenomenon. Russian scientist Alexander von Butlerov, University of Pennsylvania professor Robert Hare, English biologist Alfred Russel Wallace, and Nobel Prize winner Sir William Crookes were all convinced after careful observation that Home was truly defying gravity.

For his part, Home was unable to explain his own levitations, which he discussed in his autobiography. "I have not, and never had, the slightest power over them, either to bring them on, or to send them away, or to increase, or to lessen them. What may be the peculiar laws under which they have become developed in my person, I know no more than others. Whilst they occur I am not conscious of the mode by which they are produced, nor of the sort of manifestation that is about to occur."

Master Levitator

For all his supporters, Home also had a number of prominent detractors during his lifetime. English natural philosopher Michael Faraday, English biologist Thomas Huxley, and English scientist Sir David Brewster all attended Home séances and came away unconvinced. But it was the poet Robert Browning who became one of Home's most adamant, and vocal, opponents. Browning sat in on a Home séance and wrote to the *Times* afterward that Home was a "cheat and an impostor." Browning was so angered and disgusted that he put pen to paper to compose a long poem in 1864, "Sludge the Medium," which is basically an extended Home takedown. "I cheated when I could, rapped with my toe-joints, set sham hands at work."

Browning's hatred of Home did not, however, influence his equally famous wife, the poet Elizabeth Barrett Browning, who attended the same séance that set off her husband. On the contrary, Elizabeth became convinced of Home's abilities, which became a sore spot between the married poets, leading to serious domestic disputes (about flying mediums, mind you, not whose turn it was to do the dishes).

The Brownings, who feuded over mediums at the dinner table

How to Levitate: The Balducci Levitation Explained

By keeping your audience 8 to 10 feet away from you and carefully controlling their point of view so it remains at a constant 45 degrees, you can produce the illusion of levitating.

Start by keeping your feet close together. Then shift your weight to the foot hidden from the audience. Start slowly raising your arms. Once they are in position, the real "magic" can happen. Raise the foot that faces the audience, but only about 1 inch off the ground. Much higher and you'll spoil the illusion as people will see your other foot. Now also lift the heel of your back foot. Raise the toes as well so you're only balancing on the outside edge of your back foot, near the ball.

Congratulations! You're now levitating! You're boldly defying the laws of physics!

Careful, though—you should maintain the illusion for only about 5 to 10 seconds, or it will be difficult for your body to hold this position without wobbling. When you come back down, begin by lowering your arms. Then slowly bring your feet back down in unison. For added effect, bend your knees slightly once you touch the ground again to absorb "the impact." Illusion complete. For bonus points, make sure to throw some talking points in there about how you learned to levitate during your time studying with the fakirs of India. Cultural appropriation always helps with legitimacy.

Regardless of the veracity of his claims, Home left quite an impact on Victorian London. He suffered from tuberculosis for most of his life, and the disease finally claimed him in 1886 at age 53. In his wake, he left a debate over his abilities that continues to this day. Part of Home's legacy is that he was never actually caught perpetrating fraud in his lifetime, or at least was never publicly exposed. He was also never searched by observers before or after his séances, which means no one knew what he may have been hiding up his sleeves.

But could he actually... *fly*? No, of course not.

Home was nothing more, or nothing less, than a particularly skilled magician, the David Copperfield of his day. Like most so-called levitators, Home deployed carefully contrived optical illusions in tightly controlled environments. To help add the varnish of authenticity, Home played up the seeming spontaneity of his levitation settings, which were plotted out well ahead of time. Home may have been an early discoverer of the Balducci levitation (see sidebar on page 238), a magic trick that produces the illusion of levitating a few inches off the ground and is still used today by the magician David Blaine. Home may also have been able to increase his height at will through specific stretches and bodywork, similar to the American vaudeville performer Clarence E. Willard, who could add 7 inches vertically by "separating" his hip bones and ribs.

As for the famous out-one-window-and-in-another levitation, investigators later discovered that the gap between the iron balconies outside both windows was only about 4 feet. It was certainly feasible that Home could've created the illusion of floating across the darkened room, subsequently drifting out of one window, where he paused, before carefully leaping to the balcony of the other window. Once sufficiently composed, he could "float" back in through the second window, to the shock and awe of the séance attendees. What's more, all three eyewitnesses offered contradictory accounts of the event, which was understandable, given the dark environment.

Even though Home was bound by the same laws of gravity that keep the rest of us grounded, as an optical illusionist, his skills were inarguably impressive.

· · · · ·

A final bit of real science: The average-size adult male would need a gigantic wingspan of 22 feet to take to the air. By comparison, the bird with the largest-known wingspan, the albatross, has only 12 feet of spread. But wings are really only part of the problem (notwithstanding the practical issues of walking around with 22 feet of wings to support). Birds don't just have wings to help them fly. Evolution has given them a few other advantages as well. Birds have lightweight skeletons with hollow bones and air sacs connected to their lungs that allow for the highly efficient oxygen

One more science-defying aviation attempt doomed to failure

delivery system needed for flight. And both their wingspans and their wing muscle strength are in proportion with their body sizes.

Scientists have calculated the ratio between human size and strength and concluded that we will never be able to fly. As an organism grows, its weight increases faster than its strength, which is why 5-year-olds are better at pull-ups than 35-year-olds—because their strength-to-size ratio is higher. For a human, 22-foot wings would not only be too heavy to execute the practicalities of aerodynamics (even with the most lightweight materials); they would also be simply beyond the weight that our strength could support in the air.

The only way humans will ever fly is in our dreams. And in our machines.

Fake Moon Landings

n 2015, clips of a shocking, recently discovered interview with the famous filmmaker Stanley Kubrick began appearing online. In the video, Kubrick, who directed such iconic films as *Dr. Strangelove* and *2001: A Space Odyssey*, confessed that he had staged and filmed fake moon landings for NASA in the late 1960s. Interviewed just three days before his death in 1999, Kubrick made the interviewer sign a 15-year nondisclosure agreement. Since the NDA expired in 2014, the video could finally be published for the first time.

Kubrick: I perpetrated a huge fraud on the American public, which I am now about to detail, involving the United States government and NASA, that the moon landings were faked, that the moon landings ALL were faked, and that I was the person who filmed it.

Interviewer: Okay. (Laughs) What are you talking— You're serious. Okay.

Kubrick: I'm serious. Dead serious. Yes, it was fake.

Interviewer: Why are you telling the world? Why does the world need to know that the moon landings aren't real and you faked them?

Kubrick: I consider them to be my masterpiece.

Interviewer: And you can't take credit, or even talk about—

Kubrick: Well, I am now . . .

Interviewer: So, you can't talk to Roger Ebert about it. Does that frustrate you? Why did they have to fake it? Why would they have to do that?

Kubrick: Because it is impossible to get there . . . *2001* was very ambitious, but that's not to say that faking the moon landing was not ambitious. But I learned things from making *2001*, and that's why I got this gig in the first place.

Interviewer: That makes sense.

Kubrick: Well, it was easy for me, because I didn't think a whole lot about the morality of it. But I didn't. And I could see that Neil [Armstrong] was— he was bothered by it.

The video spreading around the internet was a clip from a soon-to-be-released movie, *Shooting Stanley Kubrick*, directed by filmmaker T. Patrick Murray. Haven't heard of him? You wouldn't be alone. Murray advertised the movie as a documentary interview with "reclusive genius Stanley Kubrick," claiming he had been granted unprecedented access to the ailing Kubrick in May 1999, based entirely on a "seven degrees of separation" connection that should have been a red flag. Kubrick rarely granted interviews. Why would he drop such a shocking reveal in an interview with a filmmaker no one had ever heard of?

Gawker posted about the film in 2015, still streamable today on Vimeo .com, declaring it to be "an absolute marvel of grating, unwatchable cinema." The film is bizarrely edited, with clips from Kubrick movies regularly interrupted by interview segments conducted in poor lighting, all loosely tied together with a disconnected voiceover from Murray describing his Kubrick obsession. Gawker reached out to Kubrick's widow, Christiane, to find out if Kubrick really had spoken with Murray. Christiane released a statement via a spokesman that "the interview is a lie, Stanley Kubrick has never been interviewed by T. Patrick Murray. The whole story is made up, fraudulent and untrue."

Stanley Kubrick displaying the trademark intensity that he would have brought to the moon landings footage

Of course, there was another little problem with the documentary, which Murray claimed to have filmed in May 1999, just three days before Kubrick's death: By May 1999, Kubrick had already been dead for two months. (Note to self: When releasing fake footage of an interview with a dead person, research their actual death date first.) The "Stanley Kubrick" in the film was actually an actor that Murray had hired to play the part.

Why did this blatantly false interview resonate with the public in 2015? Because it connected to a conspiracy theory held and perpetuated by a host of prominent personalities. It first cropped up in the immediate wake of the Apollo missions and has never really died. The theory holds that the moon landings were fake and part of a massive government-funded cover-up to win the space race. And it has had some prominent believers: podcast host Joe Rogan and InfoWars founder Alex Jones, among others.

Considering that Kubrick's groundbreaking film *2001: A Space Odyssey* includes scenes set on the moon and was released in 1968, just one year before Apollo 11's successful moon landing, it wasn't much of a stretch for

Moon Rocks

The Apollo programs collected 838 pounds of moon rocks during the six crewed missions that landed on the moon. The moon rocks are really a slam dunk for anyone needing to prove that we have been on the moon. They are easily, demonstrably distinguishable from Earth rocks and meteorites: They were clearly formed in an atmosphere without oxygen or water; they show evidence of having undergone impact events on an airless body; they have major chemical differences from any Earth rocks; and they are just a tad bit older, too—200 million years older, to be exact. Boom.

Moon rocks show distinct chemical differences from Earth rocks.

the conspiracy-minded to add a new layer: The United States government had secretly paid Kubrick to film realistic scenes of actors purporting to be astronauts walking on the moon.

Over the Moon

Fueled by massive funding, heady back-to-back scientific discoveries, and intense international competition with national security risks, the United States went from not even having a space agency in 1958 to landing a man on the moon 11 years later in 1969. It was the fulfillment of a national promise. President John F. Kennedy stood before Congress in 1961, proposing that America "should commit itself to achieving the goal, before this decade is out, of landing a man on the moon and returning him safely to the Earth." He doubled down on that pledge with a famous 1962 speech at Rice University in Houston, Texas, where he said, "We choose to go to the Moon in this decade and do the other things, not because they are easy, but because they are hard; because that goal will serve to organize and measure the best of our energies and skills, because that challenge is one that we are willing to accept, one we are unwilling to postpone, and one we intend to win, and the others, too."

Inspiring words. And they led to an enormous, almost unparalleled collective scientific accomplishment in a very short span of time. Kennedy's challenge was accepted by the American people, who wanted to establish their dominance in the international space race with the Soviet Union, thereby providing a sense of fragile security during the tensions of the Cold War. If America won the space race, it meant it could also possibly defend itself from the threat of nuclear attack by the Soviet Union. A collective breath of

Kennedy delivering his famous 1962 speech at Rice University to plenty of empty bleachers

relief was exhaled by an entire generation when Apollo 11 touched down on the moon in 1969. Neil Armstrong's famous words at that critical moment became a cultural touchpoint:

"That's one small step for man. One giant leap for mankind."

There were some, however, who had their doubts.

Bill Kaysing was a scientific technical writer with tangential connections to the space race. From 1956 until 1963, he worked as a scientific writer for Rocketdyne, the company that built the F-1 engines used in the Saturn V rockets that were employed by NASA on the Apollo missions.

"It's well documented that NASA was often badly managed and had poor quality control," he told *Wired* in 1994. "But as of 1969, we could suddenly perform manned flight upon manned flight? With complete success? It's just against all statistical odds."

By his own mysterious calculations, he estimated the odds of being able to land someone on the moon at about 0.017 percent, but he did not cite any evidence for this claim. In sharp contrast, NASA used scientific calculations and estimated the possibility of success considerably higher: 87 percent before the end of the 1960s. (NASA, it turns out, was right.)

A few other things bothered Kaysing as well. Where were the stars in the photos of the astronauts on the moon? Where was the blast crater that should be evident beneath the landing module? What was up with the way the shadows fell in the photos? Kaysing's self-published book *We Never Went to the Moon: America's Thirty Billion Dollar Swindle* came out in 1976 and laid the fertile groundwork for a series of conspiracy theories about the moon landings that have persisted to this day. Somewhere between 10 and 25 percent of Americans *still* believe the moon landings were fake, depending on what survey you cite.

Kaysing's book, full of faulty logical arguments and grainy photographs, basically boiled down to this: NASA wasn't capable of landing anyone on the moon, so it instead produced a massive cover-up. Kaysing argued that the astronauts were removed from the spaceship moments before blast-off and flown to Nevada for a quasi-vacation, where they alternated filming staged sequences of the moon landing in the Nevada desert with hitting the slots at the Las Vegas casinos before being flown to Hawaii to be put back into the capsule after the splashdown. While it took more than 400,000 NASA employees and contractors to land someone on the moon, Kaysing believed it was easier for NASA to convince nearly half a million people to keep a secret with extraordinary national and international implications than it was to produce a functioning lunar craft capable of landing on the moon. Which isn't to say the latter is easy, but can you imagine trying to prevent 400,000 people from spilling the beans?

Regardless, Kaysing's questions found a foothold—and almost all future conspiracy theories about the moon landings can be sourced to his book.

The United States of the 1970s was different from the United States of the 1960s. A certain optimism about the future of the country had faded into cynicism. The toll of the Vietnam War, the compromised gains of the Civil Rights Movement, and the corruption of the Nixon administration had all taken their toll on the American spirit. Given these factors, it became easier for some people to believe that the US government had faked the moon landings than it was to believe they had managed to produce a major scientific and engineering discovery. The reasons why were obvious to conspiracists: NASA faked the landings to avoid humiliation, to continue to receive their generous slice of the budgetary pie, and to score a much-needed PR victory against the mighty Soviet Union.

In 1980, the ever-trustworthy Flat Earth Society (see page 7) openly accused NASA of faking the moon landings, adding that "it was all directed by Stanley Kubrick." (The Flat Earth Society was not content to rest there, however. It also claimed that the whole thing was scripted by science fiction writer Arthur C. Clarke and sponsored by the Disney corporation.) The Flat Earth Society had picked up on a thread laid by Kaysing himself in his 1976 book, wherein he argued that *2001: A Space Odyssey* "had prepared the American people for filmed versions of space exploration."

Various other people and entities picked up on the claims, including Bart Sibrel, a commercial director who wrote, produced, and directed a 2001 film intended to be an exposé about the moon landings, called *A Funny Thing Happened on the Way to the Moon*. Sibrel's objections boiled down to the claim that NASA could not have achieved something in 1969 that has yet to be duplicated, 50 years later, by any other nation on the planet. Sibrel also had issues with the photographs, a recurring theme for moon landing conspiracists, and went on to claim that the Van Allen radiation belt would not allow humans to pass through it due to extreme radiation, an issue also previously raised by Kaysing in his book.

Sibrel got in the habit of approaching Apollo astronauts and asking them to put their hand on a Bible and swear that they had walked on the moon. He ambushed Buzz Aldrin with this tactic outside of a hotel in Los Angeles—and promptly got what he deserved: a punch in the face.

This response, incidentally, is more or less what these theories also deserve. So we'll let science do some counterpunching. Most of the claims made about fake moon landings amount to observations made with incomplete information or conclusions drawn by applying "Earth logic," that is, claims that are only applicable in the gravitational environment of Earth, not to the unique conditions of outer space.

Claim: *Humans can't survive leaving the Earth because of the Van Allen radiation belt.* Doing so would deliver a massive dose of radiation on anyone traveling through it, leading to radiation poisoning and death.

Saluting the stationary flag

Answer: *The NASA scientists prepared for that, and it's not quite as bad as you think.* They shielded the astronauts by building the spacecraft with aluminum hulls and chose a trajectory path intended to lessen radiation exposure. Also, they ensured the spacecraft didn't linger in the radiation belt. The fast speeds of space travel meant that by the time the astronauts reached the other side of the belt, on average each astronaut would have received a radiation dose of a little less than 1 rem (10 millisieverts). This is about the same amount of radiation someone would receive by living at sea level for three years. Relatively high for such a short period of exposure, but certainly not life-threatening.

Claim: *No one could have filmed Neil Armstrong stepping onto the moon if he was the first person to do so.*

Answer: *The lunar module itself could have.* The module had an Apollo TV camera mounted on it to give an exterior view and broadcast the signal back to Earth so that the 600 million people watching could see the event live with just a tiny delay.

Claim: *There are no stars in the photos.* NASA removed them from the photos because amateur astronomers on Earth would be able to identify the stars and what position they were observed from, thus being able to prove they were not taken from the moon.

Answer: *The photos were taken during lunar daytime.* The dim stars were outshone by the sun and its light reflecting off the moon's surface. To compensate for the lighting conditions on the moon the Apollo cameras were set to daytime exposure, with fast shutter speeds and small apertures. In those conditions, faint objects like the stars during lunar daytime are not going to appear in shots.

Claim: *The flag is fluttering in the photos, but there is not any wind on the Moon.* This means the landings were faked and actually shot on Earth.

Answer: *The flag was attached to a rod shaped like a reverse L to provide some structure so it would not droop downward.* The flag appeared to flutter when the astronauts moved it into position, but this was because without any air drag, moving the flag caused its corner to swing like a pendulum. And

If you look closely, you can see the L-shaped rod the wrinkled flag was attached to.

as for those ripples? Those were caused by the flag having been folded in storage for four days on the flight to the moon. Those were wrinkles, not ripples.

Claim: *There are inconsistencies in the photos with the light and the angle of shadows.* Artificial lights must have been used, proving the photos were shot in a studio.

Answer: *Well, yes, in the lunar environment there certainly would be inconsistencies.* There were several light sources competing with each other: the sun itself, the sunlight reflected from the Earth, the sunlight reflected from the moon's surface, and the sunlight reflected from the astronauts and the lunar module. Furthermore, lunar dust scattered the light in many directions. It would have been a studio photographer's nightmare. So, by Earth standards, there are inconsistencies with the light; however, they are to be expected in that particular environment.

Moon landing action shot

Claim: *There should be a clearly visible blast crater from the lunar module landing.*

Answer: *It's there; it's just a lot smaller than you would expect on Earth.* It was only about 4 to 6 inches, so small it's hard to see in the photographs. By the time the lunar module landed on the moon, its descent propulsion system was throttled very far down. In other words, it wasn't dropping quickly from the sky, but slowly landing. The descent engine

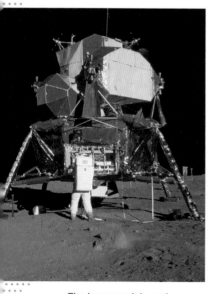

The lunar module and its tiny blast crater

The Missing Tapes

Apollo 11 telecasted SSTV (slow-scan television) in its raw format on telemetry data tape of the moon landing in 1969. These data tapes were used to record all transmitted video for backup. Meanwhile, in real time as the moon landing happened, NASA personnel converted the SSTV transmission to NTSC television format for broadcast around the world. This was all fine in theory, except for one major problem: The conversion to NTSC format for television broadcasting resulted in significant degradation of the image quality.

And as the real-time NTSC broadcast worked and was recorded widely on videotapes, preservation of the backup SSTV video was not deemed a priority. The tapes soon went missing.

The missing tapes were a real shame for the historic preservation of the significant event. Later photographs of the SSTV recordings revealed just how dramatic the difference in image quality between the SSTV recordings and the NTSC broadcast was.

But it wasn't just historic preservation that took a blow with the missing tapes. They also became fodder for the fake moon landings conspiracy theories.

Believers like to cite the missing tapes as evidence that the moon landings never actually happened. Their disappearance all seemed a little too convenient—the high-quality images of the moon landing somehow lost to time.

In 2009, NASA itself investigated the missing tapes and issued a self-damning report that the government agency couldn't find them. Worse, they were pretty sure they had recorded over them. Yep, the billion-dollar federal government's agency responsible for space exploration had done the same thing you got mad at your baby brother for doing in the 1980s: recording over your favorite show.

A stupendously boneheaded move, surely, but not a clandestine effort.

Alan Shepard playing golf on the moon

had to support only the weight of the lander, which was further helped by the Moon's gravity and the near exhaustion of the descent propellants. The lander's exhaust gases expanded quickly beyond the landing site, which did indeed scatter a lot of fine dust. The lunar regolith, a compact layer of unconsolidated debris beneath the surface dust, makes it more or less impossible for a descent engine to blast out a crater.

* * * * *

The fake moon landings claims have been debunked by scientists and journalists over and over. But what about the agency behind the moon landings themselves? NASA issued a two-page fact sheet in 1977 shortly after the publication of Kaysing's book that bluntly and briefly responds to the claims. Its memorable opening lines: "Yes. Astronauts did land on the moon." The fact sheet was reissued again in 2001 in response to Sibrel's documentary. However, the agency clearly views the conspiracy theories as so preposterous they don't warrant a larger response. (Thankfully, NASA has better things to do.)

There is one repeatedly raised question about the moon landings that *does* have some validity: Why has no one been back to the moon since Apollo 17 in 1972? The answer is not, as the conspiracists would have it, because of the lack of science or technology. The answer is the lack of money. To make Apollo 11 happen, NASA required 4 percent of the US national budget. Today, NASA is funded at 0.4 percent. A manned mission to the moon would cost about $162 billion today—an extraordinary amount of money that is difficult to raise in the halls of Congress.

Nevertheless, the Artemis program, a collaboration spearheaded by NASA with involvement from partner country space agencies and private companies, launched during the Trump administration in 2017 to reestablish a human presence on the moon. The initiative has the lofty goals of not only putting people on the moon again but also constructing a permanent base camp there to allow for future exploration. What's more, the Artemis program vows to finally do two things that the Apollo missions never did: put a woman and a person of color on the moon.

And that would indeed be a giant step for humankind.

Climate Change Denial

The year 2023 almost by itself was evidence that something was amiss with the Earth's climate.

Here's but a sampling of 2023's extreme weather: In January, a record number of storms deluged California, previously bone-dry from a years-long drought. In February, ice storms hit Texas, while the Midwest was lashed with out-of-season tornadoes. In March, tropical cyclone Freddy was one of the most long-lived on record, killing hundreds in Malawi and Mozambique. A deadly bomb cyclone hit California that same month, drenching the area yet again. In April, a bizarre "two bull's-eye" storm killed 32 people in the Midwest, South, and Southeast US. Areas of Southeast Asia baked amid a record-breaking temperature of 113°F (45°C). In May, wildfires in Canada blanketed New York City in an orange fog, making outdoor air quality awful and irritating for days. In June, the global surface air temperature was the warmest on record for that month, dating back to the seventeenth century, when we started recording temperatures.

And that's only *half* of a recent year. According to the National Oceanic and Atmospheric Administration (NOAA), the last few years (pre-2024) have been the warmest on record for the planet. So something terrible is definitely happening, and it's no longer a surprise to anyone.

It's climate change.

Every day, news organizations inform us of extreme temperatures, precipitation, wildfires, tropical storms, bomb cyclones, heat domes, and flooding, to the point where it's not really news anymore—it's just a daily disturbing reminder that our climate has warmed, causing the weather to go haywire. Because of all this, you might be tempted to think climate change is no longer debatable. But for some, it still is.

Fossil Fuel–Burning Love

Despite what some might think, the concept of climate change and the well-known atmospheric greenhouse effect has been around since the early 1800s. In 1824, French physicist and mathematician Jean-Baptiste Joseph Fourier thought the Earth should be colder given the amount of heat received from the sun, leading him to theorize that gases in the atmosphere trapped heat closer to the Earth's surface. By 1896, Swedish scientist Svante Arrhenius calculated that increased carbon dioxide (CO_2) from coal combustion could result in global warming and possibly the extinction of the human race.

Jean-Baptiste Joseph Fourier, the man who first considered the planetary greenhouse effect, looks rather sorry about our current predicament.

By the 1940s and '50s, there was a huge uptick in research on the subject, with data documenting the rising temperatures across the globe from fossil fuel burning. In 1979, the first comprehensive assessment of global

• CLIMATE CHANGE DENIAL | 255

A polar bear wonders what happened to its rapidly shrinking polar ice.

climate change due to CO_2 was released by climate scientists. Known as the Charney Report, the assessment concluded that "the most probable global warming for a doubling of CO_2 [is estimated] to be near 3°C with a probable error of ± 1.5°." So far, the calculation is on track. Atmospheric CO_2 levels measured at the Mauna Loa Observatory in Hawai'i have increased by 50 percent compared to the year 1750, and our current global temperature is about 1.2°C (2.2°F) higher than it was 100 years ago. Evidently, Earth is hotter today than it has been for the last 1,000 years.

Meaning: We're not just on our way to global warming—we're already here and are now continuing on a trajectory that we're all feeling in increasingly intense ways, in every corner of the planet.

But (there's always a but, right?) despite this clear evidence and the massive number of catastrophic weather events, the melting ice sheets, chunks of glaciers falling in the Arctic, and sad visions of polar bears trying to survive as their frozen habitat literally disappears in front of them, people still don't believe climate change is happening. According to a Pew Research survey in 2023, only 54 percent of Americans believe that climate change is a serious threat.

Why? There are a vast number of arguments that claim climate change isn't real, that it's a conspiracy theory, one pushed by a bunch of tree-huggers bent on destroying the livelihoods of hardworking coal miners and others who depend on the fossil fuel industry. But many of the questions coming from climate change skeptics in fact sound reasonable, so let's dive in.

For starters: Is there disagreement from actual scientists that climate change is real? Yes, there is. But if you're waiting for 100 percent of all scientists to agree on anything except the fundamental tenets of science, like that the Earth is round and not flat (see Flat Earth chapter, page 2), then you'll be waiting a long time. If this fact is the nugget of discord that keeps you from thinking climate change exists, consider that: Currently, 99 percent of studies show that burning fossil fuels has caused climate change. A whopping 97 percent of climate scientists who publish also agree on this. Keep in mind as well that there are a handful of junk studies by scientists funded by the fossil fuel industry, which has created slick marketing campaigns to sow doubt about climate change in the public.

No. Just no.

One of the main contributors to climate change denialism is all-or-none thinking. For climate change deniers, if all the scientists can't agree then *none* of the data is valid. But that's not how scientific consensus works—it is, however, a fantastic way to fuel skepticism. Imagine you're taking a driving test and you score a 99/100 on your written exam. In the thinking of climate change skeptics, you've failed completely and can never drive a car. It plain doesn't make sense.

What does make sense, kind of, is the skepticism stoked by the fossil fuel industry. It may be hard to believe, but there was once fervent debate around whether or not tobacco was actually dangerous for your health. Enter the "tobacco industry playbook," a well-known strategy of disinformation paid for by industry players who stood to make money by keeping tobacco (or, in the case of climate change, fossil fuels) available and selling. The classic strategy of the tobacco industry playbook is five-pronged:

1. Undermining legitimate scientific consensus by paying scientists to introduce junk research;

2. "Astroturfing," or hiding the true creator of a message so it appears to come from the general public, when really it's an industry plant working to sway the narrative in the companies' favor;

3. Lobbying and political action, which is once again paid for by the industry and which has an enormous effect on people's lives;

4. Emphasizing personal freedom versus public responsibility. The industry injects beliefs like "Smoking may be bad, but let's leave it up to the grown-up consumer to decide what's right for them. Let's not let Big Government get involved in people's personal choices";

5. Creating what appear to be nonprofit organizations that sound very legitimate but work hard at reinforcing the industry's message through publicity campaigns, more junk research, and funding of political players. Usually, these "nonprofits" are funded by those who make money by maintaining the status quo.

Contrails and Chemtrails

The sight of jets flying high above the Earth trailing whispery white cloudlike streaks is a daily occurrence. Basic chemistry tells us that when you burn airplane fuel (a mixture of petroleum-based hydrocarbons) and oxygen, the result is carbon dioxide and water. And when that leaves a jet engine at high altitude with temperatures around −34°F (−36°C), that water condenses into a cloudy vapor or ice crystals. Yes, there are small amounts of unburned fuel, sulfur compounds, and other impurities left over from the fuel, but what you see are trails of water condensation.

Some people think that these trails are much more sinister, that they're being spewed across the world in a covert plan to induce mind control, test bioweapons, or make people sick so drug companies can profit. Others think they're simply toxic to humans and wreaking havoc in less purposeful ways. Proponents of chemtrails believe that chemicals like barium, silicon carbide, and aluminum salts have been released, though expert consensus by atmospheric and geochemical scientists debunks this. Photos of Airbus and Boeing test airplanes with fluid-filled ballast tanks (to simulate the passenger weight of a full or partially full flight) are used to support the theory that harmful "chemtrails" are a thing.

If you're waiting for a whistleblower to emerge with the truth, at least listen to Edward Snowden, one of the most famous government whistleblowers in our history, who once had access to plenty of secret government filings: "In case you were wondering: Yes, man really did land on the moon. Climate change is real. Chemtrails are not a thing."

Water vapor or mind control?

Sadly, this playbook works. If it weren't for whistleblowers in the tobacco industry—and the playbook outlined in countless tobacco industry documents—we might not recognize that the fossil fuel industry is using these exact tactics to protect its own interests and undermine the Environmental Protection Agency (EPA). We know that fossil fuel companies were aware that their fuel combustion could result in greenhouse gases and result in a global crisis decades ago, yet they still poured money into organizations that sowed doubt among the public that climate change is a real threat. We also know that coal companies once told the public that more CO_2 in the atmosphere was better for plant life, which uses CO_2 to make their own energy. It isn't, of course, because of pesky things like droughts and floods. Research also shows that rice, a major grain feeding the world's population, is less nutritious when grown in higher CO_2 environments. Too much of a "good thing" for plants can be to their detriment.

Weather and Climate: Not the Same Thing

Climate skeptics will often use the existence of brutal winter storms to decry global warming, especially if they happen in warmer states. "If the world is getting so damn hot," they typically say, "then why is there a blizzard in my town right now?" Oklahoma Senator James Inhofe famously tossed a snowball in the Senate in an attempt to prove that climate change wasn't happening. It sure made for a dramatic scene, but *weather and climate are not the same thing.*

Weather is how the atmosphere acts in a short period of time today, tomorrow, in an hour. Climate is how the atmosphere behaves over a much longer span of time. But climate nonetheless affects weather every single day. For example, those blizzards that have been weirdly showing up in Texas or historically warm areas? Part of the cause could be that warm air can hold a lot more moisture than cold air, resulting in massive amounts of snow and rain. That warmer air can also have profound effects on the jet stream, which can then deliver blizzards farther south. But if you look at the larger trends—not just today's weather—winters are getting warmer everywhere by up to 2.5°F (−16°C), and there are still more record hot days than cold days across the globe.

Two degrees doesn't sound like much, does it? I mean, how bad can 2 degrees be when on a normal day, the temperature can go from 60°F (16°C) to 90°F (32°C) and back again at night? People from lots of places around the US like to joke, "If you don't like the weather, just wait a minute and it'll change!" So who really cares about 2 degrees? But remember we're talking climate, not weather. A 2-degree increase in climate change terms means melting polar ice, extreme weather happening more frequently (hot weather in Siberia!), too many American towns hitting 120°F and higher in the summer, and endangered habitats. It means creatures that like hot weather (like mosquitoes) will move north and creatures that like cold weather will find their worlds shrinking. It's about deserts expanding and northern permafrost melting and unleashing who knows what kinds of frozen viruses into the world. And 2 degrees is just an average. Some areas, such as the Arctic, have warmed a whopping 5 degrees. If the climate warms by 5.7°F, or 3°C, in this century, that would mean the death of coral reefs and the flooding of coastal American cities like Charleston, South Carolina, and Providence, Rhode Island. The National Mall in Washington, DC, would be underwater. It would mean once-in-a-lifetime floods would become regular occurrences. There would be disasters upon more disasters. So yes, just a few degrees is a really, really big deal.

US Senator James Inhofe from Oklahoma holds proof that snow exists.

But what about ice ages of the past, you might ask? Those were massive temperature changes that had nothing to do with humans or coal-fired factories. How do we really know this current warming trend is actually caused by humans and not just a geological "blip" in the grand scheme of things, like the ice ages of the past? Well, because we've been studying the big patterns of Earth's climate history for a while via climate proxies—trapped bubbles of air in ice core samples, sediment deposits, and tree rings. We know that ice ages, which dropped the Earth's temperatures by 11°F (6°C), happened because of the Earth's orbit cycles on the order of tens of thousands of years. Put simply, we get a little farther from the sun every once in a while, and an ice age happens. The CO_2 levels during those post–ice age rewarming periods are nothing like they are today—they're nearly double the amount in normal times, and the rate of the CO_2 increase is also much faster. We know the CO_2 levels of the past and can estimate the corresponding temperatures. It turns out that, with the exception of a few small wobbles year to year or within decades over the last several centuries, the temperatures have risen like a steep slope *only* in the last 150 years.

People might be tempted to consider that since climates have always changed—an ice age here, some warming up later—this trend isn't so different. Except that it is. This is yet another case of all-or-none thinking that ignores the present data, shooed away by the idea that Earth's climate changes on geological scales. Skeptics may ask, "Who are we to think our insignificant human selves are causing it?"

Well, CO_2 levels are rising so far above anything that has happened on previous geological scales that it's impossible to discount that fact. We know that volcanic activity can cool a planet with its dust or warm it by spewing methane and other greenhouse gases into the atmosphere. But volcanic activity is more likely to cool the planet than to warm it up. Then there's the sun, which seems like a steady source of light and heat, but solar irradiance actually changes on an 11-year cycle, and there are extended stretches of cooler or warmer years seen on a longer geological timeline. In the last several decades, however, solar irradiance has actually gotten lower as global temperatures have risen.

It turns out this is one of those instances when narcissism isn't imaginary. We really are the cause of climate change. We really are that powerful.

Now We're Cooking with (Greenhouse) Gas

Let's end at the beginning, with the greenhouse effect and those greenhouse gases, which is how we got into this huge mess to begin with. To maintain life on Earth, we do need *some* greenhouse gases, which include water vapor, carbon dioxide, nitrous oxide, methane, and ozone. Without them, the Earth would be too cold for our lush variety of life. The natural greenhouse

262 | PSEUDOSCIENCE •

Cows, actively farting their methane contribution to global warming

effect happens when the sun's energy penetrates our atmosphere and passes through transparent greenhouse gases like methane and carbon dioxide. That energy then hits the Earth, warming it and creating infrared radiation, and this infrared radiation is released upward. In this form, the greenhouse gases absorb some of that heat and emit it in all directions, including back down to the surface of the Earth. It's just like a glass greenhouse, in which sunlight passes through the glass and the heat stays in, bouncing off the glass and allowing you to grow tropical flowers or tomatoes in the middle of a frigid December. The concentrations of CO_2 are higher now than they've been for the last 2 million years, and the concentrations of methane are higher than the last 800,000 years—which means the greenhouse effect is intensifying in an extreme way.

CO_2, Earth's most important greenhouse gas, comes from combustion of fossil fuels, but also from deforestation and industries that create fertilizer and steel. From carbon dating, we know the CO_2 in our atmosphere comes from ancient fossil fuel sources and not newer ones like plants. (It's gotten to the point where there is so much ancient CO_2 in the air and incorporated into everything that it's messing up scientists' ability to do proper carbon dating.) Methane comes from manure and cows, but also from landfills and the fossil fuel industry. The ocean absorbs a huge amount of CO_2—20 to 30 percent of what we've made in the last decades—but it can hold only so much. The CO_2 absorbed has already increased ocean acidity by 30 percent, with notably ill effects. Some creatures with shells are finding them dissolving in the acidic water. Forests that would have been able to fix massive amounts of CO_2 are being cut down. Oddly enough, aerosol pollution—a subset of air pollution that includes sulfates, clothing dryer exhaust, and microscopic particles of soot all suspended in the air—used to cool the planet by blocking the sun, dampening the effects of the greenhouse effect. But as pollution control has improved, climate change is becoming even more apparent than before. [Disclaimer: This doesn't mean that you should actively pollute the Earth—that will do more harm than good!]

• • • • •

Thankfully, things have changed for the better around climate change denial in the last few years. At the United Nations Conference of the Parties (COP) climate summit in 2023, millions of dollars were pledged to help developing countries combat global warming effects, and countries were urged to actually transition away from fossil fuels, which hold only marked a huge change from the past. Though global 2023 carbon emissions went up, in one of the biggest fossil fuel–using countries, the US, it declined by 3 percent due to more reliance on renewable energy sources and electric vehicles, among other reasons. Compared to a decade ago, about 10 percent more Americans now believe climate change is a significant issue, up from 44 percent. And though 54 percent is a majority, it also means that nearly

On February 1, 2023, ice from freezing rain covered areas of central Texas and caused massive power outages.

as many Americans think climate change isn't a big deal or isn't real at all. Most people nevertheless believe that more must be done to combat it.

And there's perhaps the bigger problem that lurks behind the "is there" or "isn't there" issue of climate change. It's the second stage of grief: denial. Even though many people might know and believe that climate change *does* exist and *is* a problem, that doesn't mean they'll do anything about it. First, let's not forget that the largest drivers of carbon emission are forces out of the control of the individual, like corporate industry, agriculture, transportation, and fossil fuel companies—and there are plenty of powerful people in the world who want to keep it this way. Meanwhile, it would also require a change to many peoples' very comfortable lifestyles, or perhaps they are already too burdened by other enormous stressors in their lives, such as dealing with chronic illnesses and poverty. But there are steps just about anyone can take, such as voting to improve renewable energy policies, composting, using EV or hybrid vehicles, buying less stuff in general, turning your thermostat up 2 degrees in the summer and down 2 degrees in the winter, reducing food waste, weatherizing your home, choosing slower or grouped shipments when buying online, and unplugging your devices when they're not in use. The list is long—and doable.

Even with extreme weather affecting more and more folks every day, climate change denial and skepticism will undoubtedly continue. With each passing week, the news seems to bring more bad news of another climate-related disaster happening somewhere. If it affects you directly, you might alter your thinking about climate change. If it doesn't, you might shrug and wait for more evidence that the world is on fire. But, as the writers of the 1979 Charney Report chillingly put it, "A wait-and-see policy may mean waiting until it is too late."

Lawsonomy

The man who invented the passenger airliner, Alfred W. Lawson, didn't believe in energy. Literally—he didn't believe it existed at all. According to Lawson, there was neither energy nor empty space in the cosmos, just substances of varying densities. He thought that the denser substances moved toward the less-dense substances by the operation of two basic principles critical to underlying his theory of the cosmos: suction and pressure.

Lawson developed his theory of suction and pressure when he was a mere child of four years old and observed that by blowing on dust you can

266 | PSEUDOSCIENCE

The first airline passenger, and the first passengers of the airline. Plus Lawson.

move it away from you, whereas inhaling near dust will bring it closer. (These were the days before they gave kids iPads.) This observation blew his four-year-old mind, and for the rest of his life Lawson applied the theory universally to any observed phenomenon around him: We can only see because our eyes draw in light by suction. (Light is perfectly fine at entering the eye through the cornea, no suction required.) We can hear because our ears draw in sound. (Sound waves enter the outer ear all on their own.) And the force that scientists identify as gravity is really the pull of the Earth's suction.

Reality, of course, is a little different. According to Newton's law of gravitation, any two objects in the universe exert a force of attraction on each other. The greater the mass of the two objects and the shorter the distance between them, the stronger the pull of the gravitational forces. Einstein expanded on this understanding of gravity by demonstrating that gravity is not just a force; it is also a curvature in the space-time continuum. The mass of an object causes space around it to bend and curve.

Lawson's misunderstanding of gravity somehow was not a barrier to his ability to invent a machine capable of defying the very force he didn't think existed: the passenger airliner. But of course Lawson was, according to Lawson, significantly more advanced than the rest of us. "His mind responds to every question and the problems that stagger the so-called wise men are as kindergarten stuff to him," he wrote about himself, in the third person. (Never a cause for alarm.) He would later start his own university to help the rest of us catch up to him.

According to the preface to *Manlife*, a collection of Lawson's essays, "to try to write a sketch of the life and works of Alfred W. Lawson in a few pages is like trying to restrict space itself. It cannot be done."

Nevertheless, we're going to give it a try.

Flight . . . Without Gravity

Alfred William Lawson was born on March 24, 1869, in London, which of course "was the most momentous occurrence since the birth of mankind," according to Cy Q. Faunce in the preface to Lawson's autobiography. Shortly afterward, his family emigrated to Canada and then moved to Detroit, where Lawson worked as a newsboy before running away from home and spending several years riding freight cars around the country. He then took up professional baseball, playing for a variety of minor league teams between 1888 and 1895, including a brief stint pitching in the major leagues for the Boston Beaneaters (who eventually became the Atlanta Braves) and the Pittsburgh Alleghenys (who would become the Pittsburgh Pirates). Lawson then became the coach and manager of teams in the Pennsylvania State League until 1907. After the quick demise of a new professional baseball league Lawson started in 1907 to rival the American and National Leagues, he left baseball behind for the budding allure of aviation.

After founding and writing for several early aviation magazines, Lawson started designing aircraft for the army. (Incidentally, he also invented that term "aircraft," and he introduced it to Webster's dictionary when he edited its aviation section. True story.) In 1919, he invented, designed, and built the world's first passenger airliner, which was capable of holding 18 people. Lawson overcame doubts about its ability to fly by piloting it himself from Milwaukee to Washington, DC, and back again. He soon established the Lawson Aircraft Corporation in Green Bay, Wisconsin, and set about designing and building a 26-passenger airliner called the Midnight Liner.

Unfortunately, during its initial takeoff attempt, the plane never reached flying speed and crashed at the end of a field. Lawson, and the four crew members onboard, escaped without injury. With the demise of the Midnight Liner, Lawson's financial backers fled and the company went under. Despite harboring some wildly incorrect ideas about how physics worked, Lawson was considered a leading aviation thinker up until this moment. With trouble securing additional financial backing, however, Lawson was left with some time on his hands. So much time, in fact, that he developed his own theory of, well, everything, called "Lawsonomy." And that's when he really went off the deep end.

Say No to Banks, Yes to Valueless Money!

Smarting from financial woes, Lawson next surfaced in 1931 when he established the Detroit-based "Humanity Benefactor Foundation," which served as a front for an economic reform movement he started calling the Direct Credits Society. The Society declared banks the source of all evil. By milking interest from everybody they loaned money to, Lawson claimed, banks were the perpetrators of all financial woes. Lawson's timing was regrettably brilliant. The devastating economic fallout of the Great Depression created large swaths of people willing to listen to anyone who spoke about the evil of banks.

Lawson, who never came up with an idea he couldn't name after himself, proposed the "Lawson Money System" as a new economic model, but it was less a model and more a sustained rant on capitalism. In it, the gold standard was to be abolished and "valueless money" was to be issued but, curiously, not redeemable for anything. (He would have loved crypto.) All interest on debt would be abolished as well. Banks were essentially to be

LAWSONOMY

ALL OF THE CREATOR'S OBJECTIVES REQUIRE TIME, PATIENCE AND CONTINUOUS PHYSICAL, MENTAL AND MORAL STRENGTH TO ACQUIRE. FROM A SMALL BEGINNING MAN HAS CLIMBED TO GREAT HEIGHTS.

PRIOR TO THE YEAR OF

2000 — ALL RACES WILL ACCEPT LAWSONOMY PRINCIPLES

1948 — ESTABLISHED LAWSONIAN IMMUTABLE RELIGION

1943 — Established THE UNIVERSITY OF LAWSONOMY

1941 — WROTE THE BOOK "NEW SPECIES"

1939 — Revealed God's Eternal Method MAGNIFICATION (LAWSON'S BOOK, "THE ALMIGHTY")

1938 — Discovered Six Dimensions

1934 — Established BENEFACTOR

1931 — RENOUNCED HIS OWNERSHIP OF MONEY AND PROPERTY

1923 — Wrote Manlife, Creation

1922 — PENETRABILITY

1917-18-19-20-21 BUILT First AIRLINERS

1908 — Established Aircraft Industry

1904 Published Book BORN AGAIN

LAWSON LOOKING FORWARD

THERE HE STOOD, THROUGHOUT THE AGE OF EXTREME FALSITY, LIKE AN IMMOVABLE ROCK OF RIGHTEOUSNESS, FOR THE IMPROVEMENT OF MAN, BEFORE GOD

MARCH 24, 1869; BIRTH OF LAWSONOMY

Man can reach great heights by constantly looking forward and upward. Or: he can sink to horrible depths by gradually slipping down into the mire below. Be unselfish and constructive.

#goals

superseded by the government, which would replace banks as the provider of loans to individuals and businesses.

Because banks during the Great Depression were anathema to the general public, tens of thousands of people began following the Direct Credits Society, which issued its own newspaper, the *Benefactor*, and held parades and mass meetings to disseminate their ideas. It all culminated with a massive event in Detroit on October 1, 1933, complete with elaborate floats in a parade that led everyone to the Olympia Stadium, where Lawson spoke, after a 15-minute standing ovation, to a crowd of 16,000 people.

Songs were sung, such as "Hark to Lawson":

Wisely we now are living; Learning Lawsonomy
Which our great Commander has given to Humanity
Hark to the voice of Lawson teaching Natural Laws
He proves his every statement, shows to us the cause
Since Lawson gave to us the truth as in the Almighty,
We have become intelligent in Spirituality.

And "God's Gift to Man:"

God, Ruler of the Universe, has sent the first Knowlegian
To teach Humanity on earth Lawsonomy, God's flawless laws
Yes, Alfred William Lawson is God's great eternal gift to man.

As the decade carried on, Lawson's crowds began to dwindle, with that 1933 event in Detroit representing the high-water mark of popularity for the Direct Credits Society. It was time for Lawson to apply his extraordinary talents to something else.

The Gospel According to Lawson

Around this time, the University of Des Moines came up for sale. In 1942 Lawson purchased the campus, including 14 acres, 6 buildings, and dormitories capable of holding about 400 students. He soon reopened it as the University of Lawsonomy. It was defined by these tenets:

Enthused University of Lawsonomy graduate ready to spread its message

> *Education is the science of knowing TRUTH*
> *Miseducation is the art of absorbing FALSITY*
> *TRUTH is that which is, not that which ain't*
> *FALSITY is that which ain't, not that which is*

Lawson was, unsurprisingly, the self-appointed arbiter of TRUTH and FALSITY. Now with his own university established to disseminate his ideas, Lawson could really, truly build the cult in his image he so obviously craved.

Textbooks at the university, many of which were required reading before students could attend, exclusively used Lawson's writings. A basketball rule book was once discarded because Lawson hadn't written it himself. He called himself the "Supreme Head and First Knowlegian," but any teacher employed at the university could be a "Knowlegian" themself, a title that could also be claimed by any graduate. Students at least did not have to pay for their tuition or board—probably the only benefit conferred upon them by attending the University of Lawsonomy.

At the university, you could do a real deep dive into Lawson's conceptualization of physics, built on the principles of suction and pressure. As Lawson declared, "When one studies . . . Lawsonomy . . . all problems theoretically concocted in connection with Physics will fade away."

According to Lawsonomy, the human body itself was basically thousands of miniature suction and pressure pumps operating in unison. The lungs suck in air, the stomach sucks in food (with a little help, granted), and each cell contains tiny pumps that circulate blood around the body. And those things your body needs to eliminate from time to time—they leave your body thanks to pressure. By these rules, the human body is a finely tuned series of suction and pressure systems that all work in harmony... until they don't. And then you die.

Sex, too, was supposedly governed by suction and pressure. Women represented suction and men represented pressure. Sexual attraction was merely the attraction of suction for pressure.

Lawson himself never got married.

Lawson's gender theories were also able to explain the laws of magnetism. Gendered particles helped define scientific properties. So, for example, a magnet with "more female particles" had the power of suction, while a magnet with "more male particles" had the power of pressure to push matter away from it.

The Earth was described by Lawson as this suction-and-pressure theory writ large. The Earth is floating in a sea of "ether," which is a material of extremely rare density. However, within the ether is a substance of even rarer density called "lesether." Because lesether was less dense than ether, it created a suction force through a (nonexistent) hole at the North Pole. This hole sucked in various substances and gases from celestial bodies, such as meteors and the sun itself. These substances, full of nutrients for the Earth, are disseminated through a tube that extends from pole to pole through the center of Earth. From this central pole, a large number of "arteries" extend and carry these essential nutrients to all parts of the Earth, giving the planet just the right celestial nutrients to survive. The inner Earth is also crisscrossed with "veins" that flush away waste matter toward the South Pole. There, at the other opening in the Earth (yeah, Earth has an anus), waste matter could be expelled back into space.

The brain, however, was too complex for even Lawson to distill down to the concepts of suction and pressure. The brain was instead governed by tiny creatures in conflict with each other called "Menorgs" and "Disorgs." The Menorgs were the mental organizers, the good guys inside your skull.

The Disorgs were the bad guys, the vermin that "infect the cells of the mental system and destroy the mental instruments constructed and operated by the Menorgs."

For 10 blissful years, Lawson was able to teach his suction and pressure version of physics to the handful of (exclusively male) students who enrolled at the University of Lawsonomy. Despite their claims to be running a nonprofit university, however, Lawson and his financial supporters had some shady dealings that eventually caught up to them. In 1952, Lawson was hauled in front of the Senate's Small Business Committee to answer questions about what happened to the 62 war-surplus machine tools that Lawson had purchased "for educational purposes." Forty-five of them had been sold, and at a handsome profit, too. Lawson claimed to be ignorant of the details ("I never go in for figures at all"). He then launched into an explanation of Lawsonomy to the Senate committee. It was an effort that left all parties baffled and confused. As Lawson departed the meeting, he said it was "the damndest thing I've ever heard of in my life," which inspired Senator Blair Moody from Michigan to respond, "I don't know whether we're talking about the same thing, but I'm inclined to agree with you."

Perhaps a different title would make this feel more, um, profound?

It was the beginning of the end for the University of Lawsonomy. That same year, the IRS revoked the university's nonprofit status and demanded it pay back taxes. A few years earlier, the Des Moines tax assessor had inspected the site and declared it "a university in name only," saying that it would be better described as a community established for the purpose of "eulogizing Alfred Lawson."

Lawson closed the college in response, selling the land to a developer who then built a shopping mall. The story wasn't quite over, though. The university and its followers relocated to a farm near Sturtevant, Wisconsin, to contemplate their next moves. They painted "STUDY NATURAL LAW" on a big gray barn, which would become something of a landmark in the ensuing decades for motorists traveling along nearby Interstate 94.

Lawson died in 1954, two years after the official closing of his university (sadly, the suction and pressure systems in his body must have stopped working). One of Lawson's followers, Merle Hayden, converted the farm back into the university in 1957, and it, somehow, inched along until 2017 when Merle also passed away. The loyal adherent charged no admission but was reluctant to confer the degree of "knowlegian" onto anyone because, he claimed, the study of Lawsonomy was truly a lifelong endeavor. Without anyone left to carry the banner of Lawsonomy, what was left of the movement retreated underground. Soon, all traces of its physical presence would be wiped out as well.

In 2018, the nearby town of Mount Pleasant purchased the site and destroyed any remaining buildings, paving the way for a Foxconn display panel manufacturing plant. It was a sad irony that the last remnants of the dream of a man committed to eliminating banks from society would be bulldozed over by a major capitalist project.

If Lawson were still alive, he would probably have said he invented corporate bulldozing, too.

Dowsing

N **o one in the French provincial town of Château-Thierry** expected the arrival of Baroness Martine de Bertereau that day in 1629. Her carriage rumbled in from the northeast and halted at the local inn, the Fleur de Lys. The baroness, a striking, energetic woman in her early middle age, stepped out of her carriage and inquired about a room where her ill teenage son could rest for a few days before they continued their travels. Once he was safely sleeping away the remnants of his illness, the baroness unpacked her own belongings, which included a strange case full of even stranger instruments: dowsing rods.

Baroness Martine de Bertereau was famous for her ability to dowse, or locate mineral sources by following the movements of a forked stick or a special metal rod. Her employment by mining companies, royal courts, and aristocrats had sent her not only across the width and breadth of France, but throughout all of Europe itself—and far beyond. She had even dowsed in the high Andes mountains of Bolivia, seeking out gold deposits for mining companies.

In Château-Thierry, she removed one of her dowsing rods and proceeded to walk about the town holding it at chest level, her face lined with concentration. The townspeople watched, suspicious but intrigued to see this mysterious art in action. After completing her survey, the baroness returned to the central courtyard with a declaration. Beneath the ground upon which they were standing ran a natural spring of water with curative properties. She said the water coursed through deposits of gold and ferrous sulfate, an iron salt that was then called "green vitriol." She claimed that drinking this spring water would cure ailments of the liver, gallbladder, and kidneys, as well as help with dysentery and other diseases. (The spring water would not have cured these issues independently, though drinking a lot of water when you have dysentery or an infectious type of diarrhea sure helps.)

Instead of the exclamations of praise and delight that she may have expected, the townspeople greeted her proclamation with doubt. So the baroness asked for the quick formation of a committee to verify her declarations. Legal officers, apothecaries, and physicians, all of whom must have been having a slow day at work, immediately gathered together for a critical review. And, sure enough, a brief investigation proved the baroness was right. After digging in a few areas approved by the committee, the townspeople did indeed uncover a natural spring underneath the town. What's more, the water showed evidence of flowing through gold and iron deposits.

The baroness was the first woman entered into the historical record as a dowser. In her 1632 book *Explication of the True Philosophy Concerning the Primary Matter of Minerals*, cowritten with her husband, de Bertereau detailed astrological and alchemical recipes for locating mineral ores and also described the extensive 16-piece dowsing kit she carted around Europe. (Compass with seven angles? Check. Mineral astrolabe? Check. Metallic rake? Check. And of course a rotating stable of dowsing rods.)

The stick leads the way.

Saintly Dowsing

While most scholars attribute the origin of dowsing for water to Baroness Martine de Bertereau, some claim the practice predated her by a few decades. For their source, they refer to *Saint Teresa of Spain* published by Helen Hester Colvill in 1909. Colvill wrote that in 1568 Teresa was offered a site to build a convent, but there was one obstacle: no obvious water source. "But one day Antonio [a friar], standing in the church cloister with his friars, a twig in hand, made the sign of the Cross with it; 'or really,' says Teresa, 'I cannot be sure if it were even a cross; but at any rate he made some movement with the twig, and then he said, Dig just here.' They dug and lo! A plentiful fount of water gushed forth, excellent for drinking, copious for washing, and it never ran dry."

A bit of religious folklore? Or evidence that water-witching was in practice in Europe well before the baroness took up her rod? Unless further sources come to light, it's hard to say conclusively.

Enthusiasts of dowsing have claimed ancient roots for it, but the practice as we understand it today seems to have developed in Europe in the first half of the 16th century. To "dowse" is to use a forked stick, metal rod, pendulum, or other device to locate . . . *something*. That "something" is intentionally broad enough to cover basically anything a dowser wants (or is paid) to hunt for: a lost item, a hidden treasure, a buried sewer pipe, an electrical cable, an oil deposit, or even a lost pet or a missing person.

Let the Great Con(juring) Begin!

Historically, dowsing was exclusively used to locate mineral deposits. After the baroness's demonstration in Château-Thierry, word spread quickly, and dowsers everywhere started using their "abilities" to locate water sources. From there, the scope of what you could find with a dowsing rod expanded again to include pretty much everything else.

English dowser in action

The typical dowser carries a forked branch, preferably freshly cut from hazel trees (in Britain) or from witch hazel shrubs (in the United States), with one end of each Y-shaped fork held in each hand. The dowser then walks the area in question with the forked stick pointing upward. Once the dowser steps into the vicinity of the water or the mineral deposit, the butt end of the stick rotates or thrusts downward, pulled by some kind of energetic force or guided by an undefined extrasensory perception from the dowser.

The theory was that dowsing worked because metallic ores attracted certain trees, which in turn could be found to droop their branches over areas where these ores were located. Therefore, by cutting off a branch from such a tree, you could hold it and walk around waiting to see where else it lunged downward, leading to another mineral deposit. The same thinking applied to searching for water sources. Trees are attracted to water; by extension, their branches would be, too. So, if you carry around a branch, it's natural tendency will be to draw you toward water.

Divining rods seem to have first emerged in sixteenth-century Germany. At the time, German mining technology was considered the best in Europe, and German miners were in high demand. German dowsers used their rods to locate not just mineral sources but buried treasure as well. (The theory was that precious metals in buried treasure would also attract the tree branch.) The practice of hoarding treasures in underground chambers was much more common in pre-banking days, so miners with free time on their hands could pursue a lucrative side hustle trying to find it.

When German miners were imported to England, particularly in Cornwall during the Elizabethan era, they brought their dowsing rods with them. The practice of dowsing soon spread among the English population. Cornwall miners enthusiastically adopted the practice, where it persists to this day.

Cornwall miners developed a belief that the divining rod was guided by pixies or fairies. Intellectuals and priests at the time had differing views on the practice as it spread across England. Some agreed with the thinking used to justify the practice in Germany, while others condemned the practice as superstitious and fruitless. And still others considered it not just superstitious and irreligious, but outright demonic. The latter view led to some dowsers being tried for sorcery and witchcraft.

Nevertheless, the practice flourished. By the end of the seventeenth century, dowsing could be found in practically every corner of Europe. And as Europeans colonized America, Africa, and Asia, the practice of dowsing traveled with them. The mental image we have of the weather-beaten dowser walking the dry wastes of the Dust Bowl in search of a place to locate a well for a Great Plains homesteader can be directly traced to seventeenth-century Europe.

By the end of the seventeenth century, dowsing had not only spread across Europe, but the claims for what dowsing could find had expanded as well. You could of course locate water sources and mineral deposits of all types: gold, silver, lead, uranium, oil, and coal. But the list didn't stop there—you could also use your dowsing rod to locate . . . criminals.

Guy on top dowsing for minerals while other guys are actually harvesting them

Beyond a Reasonable Dowser

In 1692, a wine merchant and his wife were found in their wine cellar in Lyon, France, hacked to death with a cleaver. As the police officers searched the scene, they found some valuables missing in the couple's private residence above the wine cellar, but enough money was left behind that the police concluded the murderer had panicked and fled into the night before completing the crime. Despite their best efforts, the investigators could not find a trace of the perpetrators, and the case was quickly on the way to becoming cold when a novel suggestion was offered:

Hire a dowser.

Pressed by an angry and confused local population frightened at the possibility of a repeat crime, the police were grasping at straws. So they hired Jacques Aymar Vernay, from nearby Saint-Marcellin, who had come to some fame for his dowsing abilities. Aymar (as he was typically known) claimed to have located many springs for farmers throughout the province of Dauphiné, an alpine area well-known in France for producing dowsers. In the crowded field of late-seventeenth-century French dowsers, Vernay rose to the top.

Jacques Aymar Vernay demonstrating the smoke and mirrors effect of dowsing

Armed with his forked stick, he was deposited by the gendarmes at the scene of the crime. Where he put on quite the show. As he paced back and forth across the cellar, he began to shake and tremble nervously. But all eyes were on the forked stick in his hands, which, to the alarm of the gendarmes, twisted abruptly in his hands when he stood over the exact two places where the bodies had been found.

Like a hunting dog, Aymar had "picked up the scent" and was literally off and running. He started rapidly working his way through the streets of Lyon, following his dowsing rod as it tracked down the criminals. The rod led Aymar and the gendarmes trailing in his wake to a city gate and subsequent bridge leading over the Rhône River. On the other side, Aymar turned downstream, following the bank until he arrived at a gardener's house. He entered (trespassing was of seemingly no concern to him; the law was on his side) and found a table, three chairs, and an empty wine bottle. The dowsing rod shook violently. Aymar declared that criminals had paused here and consumed a bottle of wine. On questioning, the gardener's children revealed that that had actually happened—three strange men had arrived suddenly the day before, drunk the wine together, then took off at a run downriver.

Aymar and company were hot on the trail. Again. The group followed Aymar's rod further downriver to Beaucaire, a small village under an enormous rocky cliff. There, Aymar's rod directed them to the prison, where a cooperative guard lined up the prisoners for Aymar and the gendarmes' inspection. Sure enough, the rod dipped in front of one of the prisoners, arrested a mere hour ago for petty larceny. On questioning, the prisoner revealed that he had been involved in the murder, though he had not committed the grisly crime himself. He merely aided the true culprits by helping them carry the loot they stole from the wine cellar. The larcenist added that the murderers spoke Provençal and so were not locals to Lyon, but must have hailed from south of the country.

Armed with new information he could feed to his dowsing rod, Aymar was at it again, traveling from Lyon to Toulon on the Mediterranean coast, in the company of an entire troop of archers. Alas, they were a little too late. As Aymar's rod suspected and local observers confirmed, the fugitives had stopped the previous evening at an inn in Toulon. However, they'd boarded a small boat to carry them from the French kingdom into the comparative safety of Genoa, an Italian city. With their hopes and jurisdiction dashed at the border of France, the chase was officially called off.

The prisoner back in Beaucaire ended up paying the ultimate price for his compatriots. Instead of winning the mercy of the court for his help in directing Aymar and his divining rod, the prisoner was instead condemned to be broken at the wheel, his sentence read publicly in front of the wine shop where the murder took place.

Marine Dowsers in Vietnam

The headline on March 13, 1967, in the *Observer*, a weekly newspaper published for US forces in Vietnam, read: "Shades of Black Magic: Marines on Operation Divine for VC Tunnels." Marines on patrol in Vietnam had temporarily adopted the practice of dowsing to locate Viet Cong tunnels during Operation Independence. The article noted that the introduction of dowsing rods (two identical wires in this case) was initially "met with skepticism" but did manage "to locate a few Viet Cong tunnels."

"Private First Class Don. R. Steiner, Shadyside, Ohio, a battalion scout with 2nd Battalion, 1st Marine Regiment, tried the rods for the first time on a recent patrol. The rods spread apart as Steiner passed a Vietnamese hut. Upon checking inside the building, Marines discovered a tunnel that led to a family bunker underneath the trail, right where the rods had reacted."

Scientifically speaking, the odds of locating Viet Cong tunnels would have been about equal to random chance, but coincidences such as the one described here would have made the practice seem very appealing amid the fog of war.

The article concluded, "In this day of nuclear powered devices, it may seem that there is still room for the old, if you happen to be a believer."

It's a fabulous story, but what really happened here? Some historical detective work by R. W. Raymond conducted in the late nineteenth century revealed another layer. Aymar was already in town when the murder occurred (rather than being called in from elsewhere)—an important point, as he had gathered quite a bit of pertinent information before officially being called into the investigation. "A careful analysis of the numerous official and other records of this case shows it to be quite possible that the diviner had obtained important clues before he was publicly set to work," wrote Raymond in his report. Aymar, it turned out, was a really good detective and an even better showman, who had a keenly developed sense of what he needed to do to make extra money on the job.

Pseudoscience in Service to Science?

Scientifically speaking, dowsing is no more effective than random chance, despite many rigorous attempts to prove otherwise. The strange movements of the dowsing rod in action are attributed to the ideomotor phenomenon, wherein someone makes motions unconsciously. (This is the same effect used to explain movement on Ouija boards.) Another explanation for the phenomenon is that the dowser makes unconscious observations of the topography around them, which then influences the movement of the dowsing rod. This subconscious monitoring of the surrounding terrain demonstrates the extraordinary abilities latent in the human brain.

It's not just unconscious observation that influences the direction of the dowsing fork. At times, a dowser is consciously making choices based on careful observation of the world around them. Take Baroness Martine de Bertereau. Her discovery of the mineral waters of Château-Thierry was not the result of a forked stick suddenly pointing to a secret spring. With a keen sense of observation and scientific training beyond most women of her time, the baroness had observed that the cobblestones of the town were stained by underground currents of water rich in mineral deposits. How did she know it flowed through gold and ferrous sulfate? She paired her knowledge of geology with the powers of observation, to her own benefit.

The baroness was in fact a true intellectual force. Born in 1600, she was already fluent in three languages by the time she was married, to a baron

Depression-era dowser in the United States

20 years her senior, Jean de Chastelet. The baron had taught himself the new sciences of mineralogy and mining engineering, a body of knowledge soon absorbed by his wife. Together, they traveled the length and breadth of France, employed by the royal court to survey sites of potential mines.

Along the way, the baroness frequently used dowsing rods to point toward a mineral or water source. She preferred the public to think she was using magic in her discoveries. The alternative, that a woman could be so well trained in science as to understand mineralogy and geology, was almost inconceivable in the seventeenth century. But convincing the public that the secret of her success was magic came with risks. She was regularly accused of witchcraft and once had to flee France to lay low in Hungary for a while.

For many years, however, the scientific couple was held in high esteem by the royal court of Louis XIII. Their standing was held in significantly lower regard by the true source of power in mid-seventeenth-century France: Cardinal Richelieu, the same villain of *The Three Musketeers* fame. Richelieu was not willing to accept the baroness's scientific acumen, nor her claims of magical power as a dowser, which led him to issue an order for her and her husband's arrest. The baron was taken to the Bastille, while the baroness was held in the castle of Vincennes.

Both died in prison.

And that's how standout female scientists fared in seventeenth-century France.

Today, the baroness is viewed as the first woman mineralogist, the first woman mining engineer, and, if you put any stock in it, the first female dowser to enter the history books. An impressive legacy—but thin comfort behind the cold gray bars of a Vincennes prison.

Sources

Flat Earth:

Clifton, Edgar. *Bedford Level*. Photograph. May 11, 1904. In *Earth: A Monthly Magazine of Sense and Science* 5, no. 49–50 (August–September 1904): 1.

Doubek, James. "Daredevil 'Mad' Mike Hughes Killed in Crash of Homemade Rocket." *NPR* online. February 23, 2020.

Hunter, Dana. "Wallace's Woeful Wager: How a Founder of Modern Biology Got Suckered by Flat-Earthers." *Scientific American* Rosetta Stones. January 12, 2015.

Mirsky, Steve. "Flat Earthers: What They Believe and Why." *Scientific American* Science Talk. March 27, 2020. Podcast, MP3 audio, 33:29.

Schadewald, Robert. *The Plane Truth: A History of the Flat-Earth Movement*. 2015.

Spontaneous Human Combustion:

Arnold, Larry E. *Ablaze*. Lanham: M. Evans and Co., 1995.

Dickens, Charles. *Bleak House*. Ware: Wordsworth Editions, 1997.

Ford, Brian J. "Big Burn Theory: Why Humans Spontaneously Combust." *New Scientist* online. August 15, 2012.

Harrison, Michael. *Fire from Heaven: A Study of Spontaneous Combustion*. North Yorkshire: Methuen Drama, 1982.

Lair, Pierre-Aimé. "Classics of the Alcohol Literature: On the Combustion of the Human Body, Produced by the Long and Immoderate Use of Spirituous Liquors." *Quarterly Journal of Studies on Alcohol* 2, no. 4 (March 1942): 806–815.

———. "Classics of the Alcohol Literature: Scientific Views on the Spontaneous Combustion of Inebriates." *Quarterly Journal of Studies on Alcohol* 2, no. 4 (March 1942): 804–805.

"Literature." *Leader* town edition 2 (December 11, 1852): 1189.

Nickell, Joe. *Secrets of the Supernatural*. Amherst: Prometheus, 1988.

Ouellette, Jennifer. "Burn, Baby, Burn: Understanding the Wick Effect." *Scientific American* Cocktail Party Physics. October 12, 2011.

Rolli, Paolo Antonio. "An extract, by Mr. Paul Rolli, F. R. S. of an Italian treatise, written by the Reverend Joseph Bianchini, a prebend in the city of Verona; upon the death of the countess Cornelia Zangári & Bandi, of Ceséna. To which are subjoined accounts of the death of Jo. Hitchell, who was burned to death by lightning; and of Grace Pett at Ipswich, whose body was consumed to a Coal." *Philosophical Transactions of the Royal Society of London* 43, no. 476 (July 1745): 447–465.

Thurston, Gavin. "Spontaneous Human Combustion." *British Medical Journal* 1, no. 4041 (June 18, 1938): 1340.

SOURCES

Gasoline Pills:

Badger, Emily, and Eve Washington. "Why the Price of Gas Has Such Power over Us." *New York Times.* Updated June 20, 2023.

"Fails to Produce Naphtha from Peat." *New York Times.* December 10, 1921.

"Ford Sees Enricht About Motor Fuel." *New York Times.* April 22, 1916.

"Inventor Guilty of Gasoline Fraud." *New York Times.* November 2, 1922.

"Tells of Auto Run by Use of Chemical." *New York Times.* April 17, 1916.

"Tests a New 'Gasoline.'" *New York Times.* May 29, 1916.

Perpetual Motion Machines:

Gobert, Charles. *Philadelphia Gazette, and Daily Advertiser.* July 12, 1813.

Hicks, Clifford B. "Why Won't They Work?" *American Heritage* 12, no. 3 (April 1961): 78-98.

Inside Out. "The Unsolved Secret of David Jones' Perpetual Motion Machine." Aired October 16, 2017. BBC One. bbc.com/news/av/uk-england-tyne-41638926.

Ord-Hume, Arthur W. J. G. *Perpetual Motion: The History of an Obsession.* London: George Allen and Unwin, 1977.

Roberts, Sam. "David E. H. Jones, Scientist Whose Alter Ego Challenged Conventions, Dies at 79." *New York Times.* July 30, 2017.

World Ice Theory:

Bellamy, H. S. *Moons, Myths, and Man: A Reinterpretation.* London: Faber and Faber, 1936.

Donnelly, Ignatius. *Ragnarok: The Destruction of Atlantis.* Scotts Valley: CreateSpace Independent Publishing, 2011.

Gardner, Martin. *Fads and Fallacies in the Name of Science.* Garden City: Dover Publications, 1957.

Kurlander, Eric. *Hitler's Monsters: A Supernatural History of the Third Reich.* New Haven: Yale University Press, 2017.

———. "'One Foot in Atlantis, One in Tibet': The Roots and Legacies of Nazi Theories on Atlantis, 1890–1945." *Leidschrift: Historische Tijdschrift* 42, no. 1 (January 2017): 81–102.

Ley, Willy. "Pseudoscience in Naziland." Propagander! Accessed January 31, 2024.

———. *Watchers of the Sky: An Informal History of Astronomy from Babylon to the Space Age.* New York: Viking Press, 1963.

Wessely, Christina. "Cosmic Ice Theory—Science, Fiction and the Public, 1894–1945." Max Planck Institute for the History of Science. Accessed January 31, 2024.

Body Divination:

Anand, Akshay, et al. "Can Cheiromancy Predict Mean Survival or Fatality of a Patient with Amyotrophic Lateral Sclerosis?" *Journal of Neurosciences in Rural Practice* 11, no. 2 (April 2020): 256–260.

Carmel, Julia. "Jackie Stallone, Celebrity Astrologer and Sylvester's Mother, Dies at 98." *New York Times.* September 23, 2020.

d'Arpentigny, C. S. *The science of the hand, or, the art of recognising the tendencies of the human mind by the observation of the formations of the hands.* Translated by Edward Heron-Allen. London: Ward, Lock and Co., 1886.

Heron-Allen, Edward. *A manual of cheirosophy, being a complete practical handbook of the twin sciences of cheirognomy and cheiromancy, by means whereof the past, the present, and the future may be read in the formations of the hands, preceded by an introductory argument upon the science of cheirosophy and its claims to rank as a physical science.* London: Ward, Lock and Co., 1891.

"History of Iridology." Will Shannon. Accessed February 1, 2024.

Kohn, Livia. "A Textbook of Physiognomy: The Tradition of the 'Shenxiang Quanbian.'" *Asian Folklore Studies* 45, no. 2 (1986): 227–258.

Lucas, Teghan, Amrita Dhugga, and Maciej Henneberg. "Predicting Longevity from the Line of Life: Is It Accurate?" *Anthropological Review* 82, no. 2 (2019): 155–162.

Omura, Yoshiaki. *Acupuncture Medicine: Its Historical and Clinical Background.* Tokyo: Japan Publications, 1982.

Pack, Roger A. "Aristotle's Chiromantic Principle and Its Influence." *Transactions of the American Philological Association* 108 (1978): 121–130.

Shafer, Ellise. "Jackie Stallone, Sylvester Stallone's Mother and Celebrity Astrologist, Dies at 98." *Variety.* September 1, 2020.

Smith, Richard J. *Fortune-Tellers and Philosophers: Divination in Traditional Chinese Society.* New York: Routledge, 1992.

Stallone, Jackie. "Jackie Stallone Rumpology." *Howard Stern Show.* Broadcast live circa 2000. Daily Motion video, 55:41. dailymotion.com/video/x6xuije.

Tattersall, Nick. "Blind Psychic Gropes Buttocks to See Future." *Sounding Circle* (blog). July 9, 2002.

Phrenology:

Brontë, Charlotte. *Jane Eyre.* London: Penguin Classics, 2006.

Burrell, Brian. "The Strange Fate of Whitman's Brain." *Walt Whitman Quarterly Review* 20, no. 3 (Winter 2003): 107–133.

Collyer, Robert. *Lights and Shadows of American Life.* Boston: Redding, 1843.

Joyce, Nick, and David B. Baker. "Time Capsule: Applying Technology to Phrenology." *Monitor on Psychology* 39, no. 6 (June 2008): 22.

McCandless, Peter. "Mesmerism and Phrenology in Antebellum Charleston: 'Enough of the Marvellous.'" *Journal of Southern History* 58, no. 2 (May 1992): 199–230.

"Natural Capabilities of Negroes." *American Phrenological Journal and Miscellany* new series 8, no. 6 (June 1846): 197.

Parssinen, T. M. "Popular Science and Society: The Phrenology Movement in Early Victorian Britain." *Journal of Social History* 8, no. 1 (Autumn 1974): 1–20.

"Progress of Phrenology." *Phrenological Journal and Magazine of Moral Science* 11 (1837–1838): 342.

Simpson, Donald. "Phrenology and the

Neurosciences: Contributions of F. J. Gall and J. G. Spurzheim." *ANZ Journal of Surgery* 75, no. 6 (2005): 475–482.

Staum, Martin. "Physiognomy and Phrenology at the Paris Athenee." *Journal of the History of Ideas* 56, no. 3 (July 1995): 443–462.

Strachey, Lionel, et al., eds. *The World's Wit and Humor: An Encyclopedia of the Classic Wit and Humor of All Ages and Nations*. New York: Review of Reviews Company, 1906.

UFOlogy:

60 Minutes. Season 53, episode 35, "Navy Pilots Describe Encounters with UFOs." Reported by Bill Whitaker. Produced by Graham Messick. Aired May 16, 2021, on CBS.

Bender, Bryan. "The Pentagon's Secret Search for UFOs." Politico. December 16, 2017.

Colarossi, Jessica. "BU Astrophysicist Joins NASA Team to Study UFOs." Brink. November 21, 2022.

Cooper, Helene, Ralph Blumenthal, and Leslie Kean. "2 Navy Airmen and an Object That 'Accelerated Like Nothing I've Ever Seen.'" *New York Times*. December 16, 2017.

———. "Glowing Auras and 'Black Money': The Pentagon's Mysterious U.F.O. Program." *New York Times*. December 16, 2017.

———. "'Wow, What Is That?' Navy Pilots Report Unexplained Flying Objects." *New York Times*. May 26, 2019.

"Do Americans Believe in UFOs?" *Gallup* Short Answer. August 20, 2021.

Eghigian, Greg. "Making UFOs Make Sense: Ufology, Science, and the History of Their Mutual Mistrust." *Public Understanding of Science* 26, no. 5 (July 2017): 612–626. doi:10.1177/0963662515617706.

Gipson, Andy, Miguel Sancho, Zoë Lake, Dana Leavitt, and Ross Coulthart. "We Are Not Alone: The UFO Whistleblower Speaks." *NewsNation*. Updated June 11, 2023.

Kean, Leslie, and Ralph Blumenthal. "Intelligence Officials Say U.S. Has Retrieved Craft of Non-Human Origin." *Debrief*. June 5, 2023.

Klein, Charlotte. "Why *The New York Times*, *The Washington Post*, and Politico Didn't Publish a Seemingly Bombshell Report About UFOs." *Vanity Fair* online. June 8, 2023.

Kleinberg, Eliot. "Florida Time: The Gulf Breeze UFO Sightings." *Palm Beach Post*. October 15, 2020.

Lewis-Krause, Gideon. "How the Pentagon Started Taking U.F.O.s Seriously." *New Yorker* online. April 30, 2021.

Stieb, Matt. "The UFO Whistleblower Is Back with More Crazy Claims." *New York Magazine: Intelligencer*. June 12, 2023.

Unidentified Anomalous Phenomena Independent Study Report. NASA online. September 14, 2023.

U.S. Congress. House of Representatives House Oversight Committee, Subcommittee on National Security, the Border, and Foreign Affairs. "Unidentified Anomalous Phenomena: Implications on National Security, Public Safety, and Government Transparency." YouTube Video, 2:34:35. July 26, 2023. www.youtube.com/live/KQ7Dw-739VY?si=VjHbQi1DcoFeiWXI.

U.S. Department of Defense. Office of the Director of National Intelligence. *Fiscal Year 2023 Consolidated Annual Report on Unidentified Anomalous Phenomena.* 2023.

Wendling, Mike. "UFOs: Five Revelations from NASA's Public Meeting." *BBC* online. May 31, 2023.

The Bermuda Triangle:

Atalanta Search. *Times* (London). April 13, 1880: 6.

Barber, Elizabeth. "Bermuda Triangle Doesn't Make the Cut on List of World's Most Dangerous Oceans." *Christian Science Monitor* online. June 10, 2013.

Berlitz, Charles. *The Bermuda Triangle.* New York: Doubleday, 1974.

"The Bermuda Triangle: A Selective Bibliography." *Naval History and Heritage Command* online. January 4, 2021.

Cohen, Howard. "Coast Guard Ends Search for Missing Boat in Bermuda Triangle." *Tampa Bay Times.* January 2, 2021.

Eckert, Allan W. "The Mystery of the Lost Patrol: Five Planes Disappeared Without Trace on a Peacetime Flight." *American Legion Magazine* 72, no. 4 (April 1962): 12–13.

Gaddis, Vincent H. "The Deadly Bermuda Triangle." *Argosy* (February 1964): 28–29, 116–118.

Jones, E. V. W. "Sea's Puzzles Still Baffle Men in Pushbutton Age." *Miami Herald.* September 17, 1950.

Kusche, Lawrence David. *The Bermuda Triangle Mystery—Solved.* New York: Harper and Row, 1975.

"The Loss of Flight 19." *Naval History and Heritage Command* online. January 30, 2024.

Quasar, Gian J. *Into the Bermuda Triangle: Pursuing the Truth Behind the World's Greatest Mystery.* Camden, ME: International Marine/Ragged Mountain Press, 2005.

Sand, George X. "Sea Mystery at Our Back Door." *Fate* (October 1952): 11–17.

"Ship Deserted." *Times* (London). November 6, 1840.

Spencer, John Wallace. *Limbo of the Lost.* New York: Bantam, 1969.

"What Is the Bermuda Triangle?" *NOAA* online. January 4, 2010.

Crop Circles:

Aubrey, John. *The Natural History of Wiltshire.* London: J. B. Nichols, 1847.

Carruthers, Peter. "How Tully Became Queensland's Answer to Roswell." Produced by Gillian McNally. *Cairns Post* online. Accessed February 5, 2024.

"Crop Circles." Led Zeppelin Official Forum. August 2011. forums.ledzeppelin.com/topic/16358-crop-circles/.

Garber, Megan. "When Cameras Took Pictures of Ghosts." *Atlantic* online. October 30, 2013.

Jaroff, Leon. "It Happens in the Best Circles." *Time* online. September 23, 1991.

Moore, Patrick. "'That Wiltshire Crater' Letter to the Editor." *New Scientist* 351 (August 8, 1963).

"Mowing Devil." In *Oxford Reference.* Accessed February 5, 2024.

Perry, Michael. "Stone Wallabies Make Crop Circles." *Reuters.* June 24, 2009.

Radford, Benjamin. "Revisiting the 'Stonehenge Surprise': The 'Best Case' for Crop Circles?" *Skeptical Inquirer* 64, no. 3 (May/June 2022).

Schnabel, Jim. "Against the Grain." *Washington Post.* August 18, 1991.

Tim the Yowie Man. "Saucer Serial Hysteria: The Case of the Tully Crop Circle." *Australian Geographic* online. September 21, 2020.

Wilson, Peter. "Crop Circles Were Made by Supernatural Forces. Named Doug and Dave." *New York Times.* June 12, 2022.

Ghosts and Ghost Hunting:

1 Samuel 28:13-14 (New International Version Bible).

Ballard, Jamie. "Two in Five Americans Say Ghosts Exist and One in Five Say They've Encountered One." YouGov. October 21, 2021.

Efthimiou, C. J., and Sohang Gandhi. "Cinema Fiction vs. Physics Reality: Ghosts, Vampires and Zombies." *arXiv* physics/0608059 (September 2006). doi:10.48550/arXiv.physics/0608059.

"Less Than Half of Americans Believe Ghosts Are Real." *Ipsos* online. October 24, 2019.

Magak, Adhiambo Edith. "The Luo Dirge That Ushers the Dead to Immortality." Order of the Good Death. December 17, 2021.

Shermer, Michael. "Is it Possible to Measure Supernatural or Paranormal Phenomena?" *Scientific American* online. September 1, 2016.

Cryptozoology:

Battell, Andrew. *The Strange Adventures of Andrew Battell of Leigh, in Angola and the Adjoining Regions.* Second Series, no. 6. Edited by E. G. Ravenstein. Nendeln/Liechtenstein: Kraus Reprint Ltd., 1967.

Buhs, Joshua Blu. *Bigfoot: The Life and Times of a Legend.* Chicago: University of Chicago Press, 2009.

Darnton, John. "Loch Ness: Fiction Is Stranger Than Truth." *New York Times.* March 20, 1994.

Dendle, Peter. "Cryptozoology in the Medieval and Modern Worlds." *Folklore* 117, no. 2 (August 2006): 190–206.

Lydekker, Richard. "Gorilla." In *1911 Encyclopædia Britannica.* Last modified February 28, 2019. en.wikisource.org/wiki/1911_Encyclop%C3%A6dia_Britannica/Gorilla.

Schembri, Elise. "Cryptozoology as a Pseudoscience: Beasts in Transition." *Studies by Undergraduate Researchers at Guelph* 5, no. 1 (Fall 2011): 5–10. doi:10.21083/surg.v5i1.1341.

Shah, Shiraz A., et al. "Expanding Known Viral Diversity in the Healthy Infant Gut." *Nature Microbiology* 8 (April 2023): 986–998. doi:10.1038/s41564-023-01345-7.

2012 Phenomenon:

Andersen, Erin. "12-12-12: The End of the World or a New Beginning?" *Lincoln Journal Star.* December 7, 2012.

Baume, Maïa de la. "For End of the World, a French Peak Holds Allure." *New York Times.* January 30, 2011.

Coe, Michael D. *The Maya.* London: Thames and Hudson, 2015.

Cortes, Pepe. "Mayans Never Predicted World to End in 2012: Experts." *Reuters*. December 8, 2011.

Figueroa, Norbert. "What Did the Maya Really Say About 2012?" GloboTreks. October 28, 2022.

"Guatemalan Mayans Begin Fire Ceremony to Welcome New Era." *La Nacion*. December 21, 2012.

Makemson, Maud Worcester. *The Book of the Jaguar Priest: A Translation of the Book of Chilam Balam of Tizimin*. Chicago: Schuman, 1951.

NASA/Goddard Space Flight Center. "Magnetic Pole Reversal Happens All the (Geologic) Time." ScienceDaily. November 30, 2011.

"Pole Shift in 2003 Date." ZetaTalk. Accessed February 6, 2024.

Sinnott, Roger W. "The Great 2012 Scare and What You Need to Know." *Sky & Telescope* online. October 17, 2009.

Sitler, Robert K. "The 2012 Phenomenon Comes of Age." *Nova Religio* 16, no. 1 (August 2012): 61–87. doi:10.1525/nr.2012.16.1.61.

Smithsonian National Museum of the American Indian. "Calendar." Living Maya Time. Accessed February 6, 2024.

Waters, Frank. *Mexico Mystique: The Coming Sixth World of Consciousness*. Athens: Swallow Press, 1989.

Cryonics:

"Futurist Has Body Frozen in Hopes of Cancer Cure." *Chicago Tribune*. Updated August 21, 2021.

Johnson, Larry, and Scott Baldyga. *Frozen: My Journey into the World of Cryonics, Deception, and Death*. New York: Vanguard Press, 2009.

Markel, Howard. "How a Strange Rumor of Walt Disney's Death Became Legend." *PBS NewsHour* online. December 17, 2018.

Saver, Jeffrey L. "Time Is Brain—Quantified." *Stroke* 37 (2006): 263–266. doi:10.1161/01.STR.0000196957.55928.ab.

Shermer, Michael. "Who Are You?" *Scientific American* 317, no. 1 (July 2017): 73. doi:10.1038/scientificamerican0717-73.

Weaver, Courtney. "Inside the Weird World of Cryonics." *Financial Times* online. December 18, 2015.

Astrology:

Boland, Mary Regina, et al. "Birth Month Affects Lifetime Disease Risk: A Phenome-Wide Method." *Journal of the American Medical Informatics Association* 22, no. 5 (September 2015): 1042–1053. doi:10.1093/jamia/ocv046.

Carlson, Shawn. "A Double-Blind Test of Astrology." *Nature* 318 (December 5, 1985): 419–425.

Disanto, Guilio, et al. "Month of Birth and Thymic Output." *JAMA Neurology* 70, no. 4 (April 2013): 527–528. doi:10.1001/jamaneurol.2013.2116.

Greenbaum, Dorian Gieseler. "Astronomy, Astrology, and Medicine." In *Handbook of Archaeoastronomy and Ethnoastronomy*, edited by Clive L. N. Ruggles, 117–132. New York: Springer, 2015.

Locker, Melissa. "NASA Elegantly Shuts Down Those New Zodiac Star Theories." *Time* online. July 17, 2020.

Nederman, Cary J., and James Wray Goulding. "Popular Occultism and Critical Social Theory: Exploring Some Themes in Adorno's Critique of Astrology and the Occult." *Sociology of Religion* 42, no. 4 (Winter 1981): 325–332. doi:10.2307/3711544.

Quigley, Joan. *What Does Joan Say?: My Seven Years as White House Astrologer to Nancy and Ronald Reagan*. Secaucus: Birch Lane Press, 1990.

Tumulty, Karen. *The Triumph of Nancy Reagan*. New York: Simon and Schuster, 2021.

Ward, Kerry. "Libra Zodiac Sign: Traits and Personality Explained." *Cosmopolitan* online. December 19, 2023.

Ley Lines:

Hutton, Ronald. *The Pagan Religions of the Ancient British Isles: Their Nature and Legacy*. Hoboken: Wiley-Blackwell, 1993.

Michell, John. *The Flying Saucer Vision*. New York: Ace, 1967.

———. *The New View over Atlantis*. London: Thames and Hudson, 2001.

———. "Watkins's Revelation." *Journal of Geomancy* 3, no. 1 (October 1978): 14.

Pennick, Nigel, and Paul Devereux. *Lines on the Landscape: Leys and Other Linear Enigmas*. London: Robert Hale, 1989.

Watkins, Alfred. *Early British Trackways: Moats, Mounds, Camps and Sites*. New York: Cosimo, 2005.

———. *The Ley Hunter's Manual: A Guide to Early Tracks*. New York: HarperCollins, 1990.

———. *The Old Straight Track: Its Mounds, Beacons, Moats, Sites and Mark Stones*. London: Abacus, 1974.

Williamson, Tom, and Liz Bellamy. *Ley Lines in Question*. New York: World's Work: 1983.

Personality Psychology:

Chatterjee, Rhitu. "The Famous Big 5 Personality Test Might Not Reveal the True You." *NPR* online. July 10, 2019.

Emre, Merve. *The Personality Brokers: The Strange History of Myers-Briggs and the Birth of Personality Testing*. New York: Doubleday, 2018.

Franzen, Carl. "Talent Select AI Automatically Screens Job Candidates for Psychological and Personality Traits During Interviews." VentureBeat. June 23, 2023.

Gardner, William L., and Mark J. Martinko. "Using the Myers-Briggs Type Indicator to Study Managers: A Literature Review and Research Agenda." *Journal of Management* 22, no. 1 (February 1996): 45–83. doi:10.1177/014920639602200103.

"Homepage." Myers-Briggs Company. Accessed February 7, 2024.

Laajaj, Rachid, et al. "Challenges to Capture the Big Five Personality Traits in Non-WEIRD Populations." *Science Advances* 5, no. 7 (July 10, 2019): eaaw522.

Petticrew, Mark P., Kelley Lee, and Martin McKee. "Type A Behavior Pattern and Coronary Heart Disease: Philip Morris's 'Crown Jewel.'" *American Journal of Public Health* 102, no. 11 (November 2012): 2018–2025. doi:10.2105/AJPH.2012.300816.

Pittenger, David J. "The Utility of the Myers-Briggs Type Indicator." *Review of Educational Research* 63, no. 4 (Winter 1993): 467–488. doi:10.2307/1170497.

Sepkowitz, Kent. "Thank You for Smoking: How Big Tobacco Created the 'Type A' Personality Myth." Daily Beast. Updated July 12, 2017.

"The Story of Isabel Briggs Myers." AMT Management Performance AG. Accessed February 7, 2024.

Stricker, Lawrence J., and John Ross. "An Assessment of Some Structural Properties of the Jungian Personality Typology." *Journal of Abnormal and Social Psychology* 68, no. 1 (January 1964): 62–71. doi:10.1037/h0043580.

U.S. Equal Employment Opportunity Commission. "Employment Tests and Selection Procedures." Accessed February 7, 2024.

"William H. Sheldon, 78; Correlated Physiques and Traits of Behavior." *New York Times*. September 18, 1977.

Auguries:

Friedrich, Paul. "An Avian and Aphrodisian Reading of Homer's *Odyssey*." *American Anthropologist* New Series 99, no. 2 (June 1997): 306–320.

Gooley, Tristan. "Nature's Radar." *Journal of Navigation* (January 2006): 1–19. doi:10.1017/S0373463312000495.

Heckscher, Christopher M. "A Nearctic-Neotropical Migratory Songbird's Nesting Phenology and Clutch Size Are Predictors of Accumulated Cyclone Energy." *Scientific Reports* 8, no. 9899 (July 2018). doi:10.1038/s41598-018-28302-3.

Ingersoll, Ernest. *Birds in Legend, Fable and Folklore*. New York: Longmans, Green and Co., 1923.

Rejeesh, K. R. "Art of Parrot Predictions on Verge of Extinction." *Deccan Chronical*. October 17, 2011.

Thulaja, Naidu Ratnala. "Parrot Astrologers." Singapore Infopedia. May 22, 2002.

Polygraph:

Bittle, Jake. "Lie Detectors Have Always Been Suspect. AI Has Made the Problem Worse." *MIT Technology Review*. March 13, 2020.

Flock, Elizabeth. "NSA Whistleblower Reveals How to Beat a Polygraph Test." *U.S. News* online. September 25, 2012.

Iacono, William G., and Gershon Ben-Shakhar. "Current Status of Forensic Lie Detection with the Comparison Question Technique: An Update of the 2003 National Academy of Sciences Report on Polygraph Testing." *Law and Human Behavior* 43, no. 1 (February 2019): 86–98. doi:10.1037/lhb0000307.

Rutbeck-Goldman, Ariela. "An 'Unfair and Cruel Weapon': Consequence of Modern-Day Polygraph Use in Federal Pre-Employment Screening." *UC Irvine Law Review* 7, no. 3 (December 2017): 715–748.

Sun, Z. Y., et al. "Polygraph Accuracy of Control Question Test in Criminal Cases." *Fa Yi Xue Za Zhi* 35, no. 3 (June 2019): 295–299. doi:10.12116/j.issn.1004-5619.2019.03.006.

U.S. Congress. Office of Technology Assessment. *Scientific Validity of Polygraph Testing: A Research Review and Evaluation*. Government Printing Office, 1983.

———. U.S. Senate Committee on Labor and Human Resources. *Polygraphs in the Workplace*. 110th Cong., 1st sess., S. Hrg. 110-185. Government Printing Office, 1987.

U.S. Department of Justice. "262. Polygraphs—Introduction at Trial." *Criminal Resource Manual*. Updated May 2005.

Vittorio, Michele. "The Effectiveness and Future of Polygraph Testing." EBP Society. September 3, 2020.

Superstitions Hall of Fame:

Mandal, Fatik Baran. "Superstitions: A Culturally Transmitted Human Behavior." *International Journal of Psychology and Behavioral Sciences* 8, no. 4 (January 2018): 65-69. doi:10.5923/j.ijpbs.2018 0804.02.

Panesar, Nirmal S., et al. "Is Four a Deadly Number for the Chinese?" *Medical Journal of Australia* 179, no. 11-12 (December 2003): 656-658. doi:10.5694/j.1326-5377.2003 .tb05741.x.

San Filippo, Michael. "Why Do Italians Consider Friday the 17th Unlucky?" Thought Co. Updated November 28, 2020.

"Who Was St. Dunstan?" St. Dunstan's Episcopal Church. Accessed February 7, 2024.

Yuko, Elizabeth. "Why Black Cats Are Associated with Halloween and Bad Luck." History. Updated August 11, 2023.

Levitation:

Adare, Viscount. *Experiences in Spiritualism with D. D. Home*. N.p.: White Crow Books, 2011.

Avila, Teresa of. *The Autobiography of St. Teresa of Avila: The Life of St. Teresa of Jesus*. Translated by David Lewis. Charlotte: TAN Books, 2011.

Barton, Marcella Biro. "Saint Teresa of Avila: Did She Have Epilepsy?" *Catholic Historical Review* 68, no. 4 (October 1982): 581-598.

David-Neel, Alexandra. *Magic and Mystery in Tibet*. Mineola: Dover Publications, 1971.

Dennett, Preston. *Human Levitation: A True History and How-To Manual*. Atglen: Schiffer, 2006.

Dingwall, Eric John. *Some Human Oddities—Studies in the Queer, the Uncanny and the Fanatical*. London: Home and Van Thal, 1947.

Home, Daniel D. *Incidents in My Life*. New York: Cosimo, 2005.

Jacolliot, Louis. *Occult Science in India and Among the Ancients*. New York: Cosimo, 2005.

"The Lung-Gom-Pa Runners of Old Tibet." Outside *Trail Runner Magazine*. Updated May 12, 2022.

Nickell, Joe. "Secrets of 'The Flying Friar': Did St. Joseph of Copertino Really Levitate?" *Skeptical Inquirer* online 42, no. 4 (July/August 2018).

Radford, Benjamin. "Bringing Levitation Down to Earth." *Skeptical Inquirer* 27, no. 2 (November 24, 2017).

Richards, Steve. *Levitation: What It Is, How It Works, How to Do It*. Newburyport: Weiser Books, 2015.

Stein, Gordon. "The Lore of Levitation." *Skeptical Inquirer* 13, no. 3 (Spring 1989): 277-288.

Fake Moon Landings:

Bussey, Ben. "Stanley Kubrick Estate Rubbishes 'Confession' Video About Faking the Moon Landings." Yahoo News. December 12, 2015.

Cohen, Rich. "How Stanley Kubrick Staged the Moon Landing." *Paris Review*. July 18, 2019.

Evon, Dan. "Did Stanley Kubrick Fake the Moon Landings?" Snopes. December 11, 2015.

Godwin, Richard. "One Giant . . . Lie? Why So Many People Still Think the Moon Landings Were Faked." *Guardian*. July 10, 2019.

Hess, Amanda. "They Kinda Want to Believe Apollo 11 Was Maybe a Hoax." *New York Times*. July 1, 2019.

Kaysing, Bill. *We Never Went to the Moon: America's Thirty Billion Dollar Swindle*. Scotts Valley: CreateSpace Independent Publishing, 2017.

Keneally, Meghan. "Why the Apollo 11 Moon Landing Conspiracy Theories Have Endured Despite Being Debunked Numerous Times." *ABC News* online. July 18, 2019.

Kennedy, John F. "Address at Rice University on the Nation's Space Effort." September 12, 1962. Rice University Stadium, Houston. John F. Kennedy Presidential Library and Museum.

Launius, Roger D. "Yes, the United States Certainly DID Land Humans on the Moon." *Smithsonian Magazine* online. May 16, 2019.

"Moon Landing Conspiracy Theories, Debunked." Royal Museums Greenwich online. Accessed February 8, 2024.

Murray, T. Patrick, dir. *Shooting Stanley Kubrick*. 2015; Independent Film Studios, vimeo.com/148297544.

Ostovar, Michele. "The Decision to Go to the Moon: President John F. Kennedy's May 25, 1961, Speech Before a Joint Session of Congress." NASA online. September 22, 1998.

Climate Change Denial:

Carbon Dioxide: Vital Signs of the Planet. NASA online. Last modified January 30, 2023.

Climate Change: Vital Signs of the Planet. *NASA* online. Last modified January 30, 2023.

Gillis, Justin, and Clifford Krauss. "Exxon Mobil Investigated for Possible Climate Change Lies by New York Attorney General." *New York Times*. November 5, 2015.

Hirji, Zahra. "The World Is on Track to Warm 3 Degrees Celsius This Century. Here's What That Means." BuzzFeed News. October 30, 2021.

"June 2023, Record Warmth in Global Surface Air and Sea Surface Temperatures." Copernicus. July 11, 2023.

Kaur, Harmeet. "Edward Snowden Searched the CIA's Networks for Proof That Aliens Exist. Here's What He Found." *CNN* online. October 23, 2019.

Nicholls, Neville. "The Charney Report: 40 Years Ago, Scientists Accurately Predicted Climate Change." Phys.org. July 23, 2019.

Nuccitelli, Dana. "New Study Undercuts Favorite Climate Myth 'More CO_2 Is Good for Plants.'" *Guardian*. September 16, 2016.

Ocean Acidification. *NOAA* online. Updated April 1, 2020.

Rao, Devika. "The Extreme Weather Events of 2023." *The Week*. December 26, 2023.

Rosen, Julia. "The Science of Climate Change Explained: Facts, Evidence and Proof." *New York Times*. Updated November 6, 2021.

Shapiro, Jeremy P. "The Thinking Error That Makes People Susceptible to Climate Change Denial." *Conversation*. May 2, 2023.

Shearer, Christine, Mick West, Ken Caldeira, and Steven J. Davis. "Quantifying Expert Consensus Against the Existence of a Secret, Large-Scale Atmospheric Spraying Program." *Environmental Research Letters* 11, no. 12 (November 22, 2016): 129501.

Tyson, Alec, Cary Funk, and Brian Kennedy. "What the Data Says About Americans' Views of Climate Change." Pew Research Center. August 9, 2023.

University of Exeter and Stanford Doerr School of Sustainability. "Global Carbon Emissions from Fossil Fuels Reached Record High in 2023." Stanford Doerr School of Sustainability. December 5, 2023.

Lawsonomy:

Gardner, Martin. *Fads and Fallacies in the Name of Science*. New York: Dover Publications, 1957.

Henry, Lyell D., Jr. *Zig, Zag and Swirl: Alfred W. Lawson's Quest for Greatness*. Iowa City: University of Iowa Press, 1991.

Lawson, Alfred W. *Manlife*. Detroit: Humanity Publishing Co., 1923.

———. *Short Speeches as Spoken by Alfred Lawson*. Detroit: Humanity Publishing Co., 1942.

Dowsing:

Bird, Christopher. *The Divining Hand: The 500-Year-Old Mystery of Dowsing*. Atglen: Schiffer Publishing, 2000.

Colvill, Helen Hester. *Saint Teresa of Spain*. London: Methuen, 1909.

Ellis, Arthur Jackson. "The Divining Rod: A History of Water Witching, with a Bibliography." U.S. Geological Survey Water Supply Paper 416. Washington, DC: Government Printing Office, 1917. doi:10.3133/wsp416.

Raymond, Rossiter W. "The Divining Rod." *Journal of the Franklin Institute* 119, no. 1 (January 1885): 1–18.

"Shades of Black Magic: Marines on Operation Divine for VC Tunnels." *Observer*. March 13, 1967.

Underwood, Peter. *The Complete Book of Dowsing and Divining*. London: Rider, 1980.

Vogt, Evon Z., and Ray Hyman. *Water Witching U.S.A.* 2nd Edition. Chicago: University of Chicago Press, 2000.

Webster, Richard. *Dowsing for Beginners*. Woodbury: Llewellyn Publications, 1996.

Acknowledgments

The creation of a book like this is not possible without those who make the magic happen behind the scenes. Lydia and Nate would like to thank our amazing agents, Eric Myers, Michelle Tessler, and Jordan Hill, for having our backs. We would also like to give a special shout-out to Eric Myers, who has shepherded us through three books together and many others on Lydia's list—we are grateful for your vision and years of dedication to the literary world.

To the incredibly talented team at Workman Publishing and Hachette Book Group: John Meils, Lia Ronnen, Julia Perry, Kim Daly, Chloe Puton, Rebecca Carlisle, Allison McGeehon, plus so many others. Thanks again to Kelly Gonzalez for your bibliographical wizardry.

FROM NATE:

To my family, April Genevieve Tucholke, Tom Pedersen, Oscar "Shaun" Pedersen, and the extended Pedersen Clan of Richard, Inga, and Laura. For reading selected chapter drafts and offering feedback, James Danky of the University of Wisconsin-Madison and Dr. W. Todd Groce and Dr. Stan Deaton of Georgia Historical Society. For all the ongoing author support, and for running one of the best indie bookshops in the country, Joni Saxon-Giusti and Chris Blaker of The Book Lady Bookshop in Savannah, Georgia.

FROM LYDIA:

To my family: Bernie, Benjamin, Maia, and TG, Chang-Wuk and Kyong-Ja Kang, Yingming and Yachin Su, Alice and OhSang, Rich and Dana, Jen and Aaron—I am so grateful to be part of your lives and thankful for your endless support. To my lovely nieces and nephews, Sam, Ethan, Lauren, Elliot, Samantha, Owen, Natalie, and Garrett—thanks for tolerating your weirdo Auntie/Eemo! A special thank-you to Sarah Kosa for reading sections of the book and for your endless wisdom and friendship. Thanks to Adam Christopher and Angie Hawkins for lending me your ghost stories and your ears. To Felicity Bronzan, who keeps my world spinning, and to Sarah Weiss, who keeps the spin at a safe rotational speed. To Ann Kim and Tosca, my other sisters, big hugs. To my friends, colleagues, and patients at Nebraska Medicine—thank you from the bottom of my left ventricle. Thanks to *The Linden Review*, which published an earlier version of our Phrenology chapter. And finally, to the librarians at the University of Nebraska Leon S. McGoogan Health Sciences Library, who have yet to call the FBI on my bizarre literature requests.

Index

A

Abbagoochie, 128
Ablaze (Arnold), 20
Abominable Snowman, 128, 135
Abominable Snowman: Legend Come to Life (Sanderson), 131
abysmal gigantism, 135
acetone, **25**, 26
acetylene, 26
Achilles, **5**, 200
acquiescence bias, 195
Adare, Lord, 235–236
Adorno, Theodor, 170
Advanced Aerospace Threat Identification Program, 84
Aeschylus, 199
aetheric wind, 8
agonic lines, 98–99
Ahnenerbe, 49
AI, personality assessment and, 194
airborne clutter, 85
airliners, passenger, 265, **266**, 267, 268
al-Bīrūnī, Abū al-Rayḥān Muḥammad ibn Aḥmad, 4, 163
alcohol/alcoholics, 17–19, 58
Alcor Life Extension Foundation, 151, 157, **158**
Aldrin, Buzz, 248
alectryomancy, 51
aliens/extraterrestrial life, **viii**, 83–84, 86–89, 96–97, 176–178, **179**. *See also* UFOlogy
ALS (amyotrophic lateral sclerosis), 61
American Association of Astrologers, 166
American Museum (Barnum's), 135
American Psychological Association, 188, 189, 216, 218
Americans with Disabilities Act (ADA), 194
Amos, Sam, 53
Anthropometric Society of Philadelphia, 66, 67
Apollo program, 243–244, 246, 248–252
Aquarius (astrological sign), 165, 169
Aragoncillo, Leandro, 216
Archimedes, 82
Architeuthis dux, 133
Argosy, 95
Argüelles, José, 142
Aries (astrological sign), 165, 169
Aristophanes, 199–200
Aristotle, vii, 4, 57–58, 133
Armstrong, Neil, 242, 246, 249
Arnold, Kenneth, 79–80
Arnold, Larry E., 20
Arrhenius, Svante, 254
Artemis program, 252
Aryans, 49
ascendant signs, 165, 166
astrology, 160–170
astronomy, 162–163
astronomy of the invisible (kosmotechnische Weltanschauung), 41
astroturfing, 257
Atalanta, HMS, 93, **93**, 101
Atkinson, Richard, 182
Atlantis, 44, 47, **47**, 97, 146
atmospheric phenomena, natural, **80**, 85
Aubrey, John, 104
auguries, 197–208
authority, distrust of, 10–11
Avebury circle, 105, 107
Avenger torpedo bomber planes, 90–92, 95
Aymar Vernay, Jacques, 280–282
Ayurvedic writings, 191

B

baktuns, 139, 140–141, 148
Baldr, 225
Balducci levitation, 238, 239
ball lightning, 79, 107
Bandi, Cornelia Zangheri, 12–16
banks, Lawsonomy and, 268, 270
Barnum, P. T., 115–116, 135, 170
Barnum effect, 170, 188
baseball, Lawson and, 267
Basic Instinct, 210, **214**
Battel, Andrew, 132
beauty marks/moles, **55**, 56
beavers, 130, **130**
Bedford, James, 152
Beethoven, Ludwig van, 66
Bellamy, Hans Schindler, 42–43
Benard, Sigrid, 137, 138
Benefactor, 270
Benham, William, 60

Ben-Shakhar, Gershon, 218
Berlitz, Charles, 92, 95, 98
Bermuda Triangle, 90–101, **92**
Bermuda Triangle Mystery—Solved, The (Kusche), 100
Bermuda Triangle, The (Berlitz), 92, 95, 98
Bernhardt, Sarah, 60
bestiaries, medieval, 130–131
Bhāskara II, 37
Bhāskara's wheel, 37, **37**
Bianchini, Monsignor, 13–14
Bible, 57, 112–113, 224
Bigelow, Robert, 84
Bigfoot, 125–128, **127**, 136
birds
  augury and, 197–208
  flight and, 239–240
Birds, The (Aristophanes), 199–200
birthmarks, 55
black cats, 221–222
Blaine, David, 238
Bleak House (Dickens), 14–15
blood pressure, lie detection and, 211–212
Blount, Elizabeth, 6–7
Blumenthal, Ralph, 87
body divination, 50–63
body types, 192
Bower, Doug, 108–109, **109**
Bowes, Shauna, viii
Brady, James, 160
brains
  cryonics and, 155–156, 157
  function of, 154
  Lawsonomy and, 272–273
  postmortem study of, 66
  preserved, **154**
Brewster, David, 237
Briggs, Katharine Cook, 185–187, **186**
British Medical Journal, 54
British Patent Office, 38

Broca, Paul, 66
Broussais, François-Joseph-Victor, 70
Brownian ratchet, 39
Browning, Elizabeth Barrett, 237, **237**
Browning, Robert, 237, **237**
Buck, Ulf, 53
Bugarach, France, 137–138
Bush, George W., 216
Byron, Lord, 66

C

caddisfly, 134
Calendar Round, **139**, 140
Cambridge Analytica, 195–196, **195**
camphorated oil, **16**
camphorated spirits, 15–16
Cancer (astrological sign), 165
capillary action, 38
Capricorn (astrological sign), 165
Capron, John Rand, 104
carbon dioxide, 121–122, 254–255, 258, 259, 261–263
carbon monoxide poisoning, 121–122
Carlson, Shawn, 169
Carmelites, 229
cars
  electric, 30
  names of, 224
  nuclear-powered, 29–30, **29**
Carthaginians, 203, 205
cats, black, 221–222
Cayce, Edgar, 97, **97**
CCP, 185, 187
Central Medical College of Syracuse, 71
cereologists, 106
chakras, 56
Champ, 128

Chariots of the Gods (von Däniken), 176–177
Charles Bonnet syndrome, 121
Charney Report, 255, 264
Château-Thierry, France, 275–276, 277, 283
"chemtrails," 258
chickens, augury and, 202–203, **203**, 205–206
chirology (palmistry), 57–62
chiromancy (palmistry), 57–62
choleric personality, 191
Chorley, Dave, 108–109, **109**
chupacabra, 128
Churchward, James, 146
Cicero, 203, 206
cigarettes/tobacco industry, 192–193, **256**, 257, 259
circumnavigation, 4
Civil Rights Movement, 247
Civil War, 114–115
Clarke, Arthur C., 247
climate change denial, 253–264. *See also* weather/weather patterns
Clinton, Bill, 216
Cloud Atlas (Mitchell), 55
Coe, Michael D., 141
cold reads, 62–63
Coles, L. Stephen, 152
Collyer, Robert, 70, 72–73
Columbus, Christopher, 4
Colvill, Helen Hester, 277
comets, 47
comparison question technique (CQT), 212–214, 216, 218
compurgation, 210–211
Condon report, 80–81
confirmation bias, 170, 220
Congreve, William, 38
connectome, 153–155, 159
conspiracy theories, viii–ix, 10. *See also* moon landings, fake

constellations, 9, **164**, 165, 167–168, **168**
contrails, 258, **258**
Copernicus, 40
"Coral Triangle," 92
Cosmopolitan, 164, 183
Cosmos, 83
Covindasamy, 233
Crew, Gerald "Jerry," 125–126, **126**
Crookes, William, 236
crop circles, 102–111
crossing fingers, 225–226
Crowe, Russell, 28
crust displacement theory, 143, 146
cryonics, 150–159
cryoprotectants, 156
cryptozoology, 125–136
Cumberland Dragon, 128
Cusack, John, 143
cyanide, 22, **25**, 26
Cyclops, USS, 94, **94**, 101

D

David-Neel, Alexandra, 233–234
De animalibus (Magnus), 130
de Bertereau, Martine, 275–277, 283–284
de Chastelet, Jean, 284
De natura deorum (*On the Nature of the Gods*; Cicero), 203
De Niro, Robert, 210, **210**
death, fear of, 151–152
Debrief (website), 87
Defense Intelligence Agency, 216
dermatoglyphics, 60–61
Devil's Sea, 98
Devil's Triangle, 92, 95. *See also* Bermuda Triangle
Dewars, 152, 158
Dickens, Charles, 14–15

Dickinson, Rod, 111
dimethyl sulfoxide, 156
Direct Credits Society, 268, 270
Disney, Diane, 153
Disney, Walt, 152–153, **152**
Disney corporation, 247
"Disorgs," 272–273
divining rods. *See* dowsing/dowsing rods
domain awareness, gaps in, **80**
donkey's heads, using for prediction, 51
Donnelly, Ignatius Loyola, 47
Down syndrome, 61
dowsing/dowsing rods, 118, 275–284
dragon paths, 178
dreams, 44–45
Drepana, Battle of, 203, 205
Dunstan, St., 223

E

Early British Trackways (Watkins), 172
Earth
  axial tilt of, 4
  axis of, 169
  circumference of, 4
  as flat, 2–11
  Gaia theory and, 107
  magnetic field of, 98–99
  magnetic pole reversal and, 143, 145, 146
  as pendulum, 44–45
  photographs of, 7–8
  predicted collision involving, **141**, 143–145
  radius of, 4
  suction and pressure theory and, 272
Earth Mysteries movement, 178, 182
Earth's Shifting Crust, The (Hapgood), 146

Economist, 107
ectomorphs, 192
Edison, Thomas, 26, 60, 115
Einstein, Albert, 41, 146, 267
electric cars, 30
electrodermal activity, 212
elemental zodiacs, 165
elf circles, 104, **104**
EMF meters, 118, 119–120, **119**, 124
Emmerich, Roland, 143
Employee Polygraph Protection Act (1988), 217
"emsek el-khashab," 221
endomorphs, 192
energy
  denial of, 265–274
  thermodynamics and, 35–36
Enlightenment, vii
Enricht, Louis, 21–26
entropy states, 35
Environmental Protection Agency (EPA), 259
epilepsy, 232
Epstein, Jeffrey, 152
Equal Employment Opportunity Commission, 188
Eratosthenes, 4
Esfandiary, Fereidoun M. (FM-2030), 156–157, **157**
ether, 272
excrement, using for prediction, 51
Explication of the True Philosophy Concerning the Primacy Matter of Minerals (de Bertereau), 276
extraterrestrial life/aliens, **viii**, 83–84, 86–89, 96–97, 176–178, **179**. *See also* UFOlogy

extroverts, 186, 187
eyes
    phrenology and, 65
    using for prediction, 53–54, **54**

F

Facebook, 195–196
facial features, reading of, **69**, 70
Fads and Fallacies in the Name of Science (Gardener), 45
Faherty, Michael, 20
fairy rings, 104, **104**
fake moon landings, 241–252
Faraday, Michael, 237
Faunce, Cy Q., 267
Fauth, Philipp, 45
FBI polygraph program, 218
Feejee mermaid, **134**, 135
ferrous sulfate, 276
Feynman, Richard, 39
finger monitors, **217**
fingernails, 56, 58–59
fingerprints, 60–61, 61, **61**
fingers
    crossing, 225–226, **226**
    length of, 61
Finney, Hal, 152
Firepower International, 28
five-factor model of personality (Big Five), 193–195, **193**
flat Earth, 2–11
Flat Earth Society, 247
Flatwoods Monster, 128, 136
Flight 19, 90, 95, 100
"FLIR1," 77–78, 83–84
flood, legends of, 44
flying. *See* levitation
Flying Saucer Vision, The (Michell), 178
flying saucers. *See* UFOs/UFOlogy

Flying Saucers: A Modern Myth of Things Seen in the Sky (Jung), 80
FM-2030, 156–157, **157**
"foo fighters," 79
Ford, Christine Blasey, 210
Ford, Henry, 23–24, **23**
Ford Nucleon, 29–30, **29**
foreign adversary systems, **80**
Forer effect, 188
forward-looking infrared cameras (FLIRs), 78
fossil fuels, 256, 257, 259, 262–263
Fouke Monster Festival, 136
Fourier, Jean-Baptiste Joseph, 254, **254**
Fowler, Charlotte, 71
Fowler, Lorenzo Niles, 71
Fowler, Lydia Folger, 71
Fowler, Orson Squire, 67, 71
Fox, Maggie and Kate, 114, 115, **115**
Franch, Guido, 26–27
Francis II, Holy Roman Emperor, 68
Franklin Institute, 34
Fravor, David, 78
Freud, Sigmund, 186
friction, 37
Friday the 13th, 225
Frozen: My Journey into the World of Cryonics, Deception, and Death (Johnson), 151
Fulton, Robert, 33–34, **34**
Funny Thing Happened on the Way to the Moon, A (Sibrel), 247, 252
furcula, 222

G

Gaddis, Vincent, 95
Gaia theory, 107

Galen, 162, 191
Galileo Project, 86
Gall, Franz Joseph, 65, **65**, 68
ganzhi (sexegenary calendar), 167
Gardener, Martin, 45
Garrett, Michael, 82
gasoline pills, 21–30
gasoline prices, 21
Gauquelin, Michael, 170
Gawker, 243
Gemini (astrological sign), 165
gender theories, Lawson and, 272
General Social Survey, 166
geocaching, 175
geoglyphs, 173
Ghost Adventures, 118
Ghost Festival, 113
Ghost Hunters, 118
Ghost Nation, 118
"ghost rockets," 79
Ghostbusters, 116–118, **117**
ghosts/ghost hunting, 112–124
gigantism, abysmal, 135
Gigantopithecus, 127
"GIMBAL," 77, 78, **78**, 83–84
Gimlin, Robert, 126–127
"girdle tides," 43, 44
glacial till, 47
Glazial Kosmogonie (*Glacial Cosmogony*; Hörbiger), 45
gluteal cleft, 52
Goatman of Maryland, 128
Gobert, Charles, 32
"God's Gift to Man," 270
"GOFAST," 77, 78
gold standard, 268
Gooley, Tristan, 207
Gorbachev, Mikhail, 161–162
gorillas, 132
Grafton Monster, 128
gravity, 8, 9, 32, 42–44, 45, 48, 169, 229, 266–267

Great Depression, 268, 270
Green River Killer, 216
greenhouse effect, 254, 261–263
Gregorian calendar, 139
Griffin, Merv, 161
Groundhog Day, 222
Grusch, David Charles, 87–89, **88**
Gulf Breeze Sentinel, 76
Gulf Stream, 99

H

Haab calendar, 139–140
hair, using for prediction, 51
Hampden, John, 6
Hapgood, Charles, 146
Hare, Robert, 236
"Hark to Lawson," 270
Hayden, Merle, 274
health
   natural astrology and, 163–164
   personality types and, 192–193
health psychology, 193
Heckscher, Christopher, 208
Heron-Allen, Edward, 56
Heuvelmans, Bernard, 131
Hicks, Clifford B., 36
high blood pressure, 61
hill figures, 173–175
Hillary, Edmund, 135
Himmler, Heinrich, 49
Hinduism, 59
Hinkley, John, Jr., 160–161
Hippocrates, 162, 191
Historia Animalium (Aristotle), 57–58
Hitler, Adolf, 40, 41, 44, 49
Hodr, 224–225
Home, Daniel Dunglas, 234–239, **234**
Home Book of Verse, The, 58
Homer, 5, 200

Honey Island Swamp Monster of Louisiana, 128
Hörbiger, Hanns, 40–42, 44–46, 48–49, **48**
horoscopes, 165, 166, 170
horseshoes, lucky, 223
Hughes, Mike, 2–3, **3**, 11
human body, composition of, 19
Humanity Benefactor Foundation, 268
humoral theory, **190**, 191
Hungry Ghost Festival, 113
Hutton, Ronald, 178
Huxley, Thomas, 237

I

Iacono, William, 218
ice ages, 260–261
Ideler, J. L., 191
ideomotor phenomenon, 283
Iliad, The (Homer), 200
In Search of..., 127
Industrial Revolution, vii, 38
inertia, 37
Information Bureau (Vienna), 46
infrared radiation, 262
Inhofe, James, 259
International Academy of Astronautics, 82
International Flat Earth Research Society, 7
International Space Station, 7, 10
Into the Bermuda Triangle (Quasar), 92
introverts, 186, 187
Invisible Horizons (Gaddis), 95
Invisible Landscape, The (McKenna), 142
Invisible Residents (Sanderson), 96–97
iridology, 53–54, **54**

iron
   horseshoes and, 223
   touching, 221
Iron Age, 174

J

Jacolliot, Louis, 233
Jersey Devil, 128
Jesus, 224, **225**
Jet Propulsion Laboratory, 77
jet stream, 259
Johnson, Larry, 151
Johnston, Tim, 28
Jones, Alex, 243
Jones, David, 39, **39**
Jones, Edward Van Winkle, 95
Joseph of Cupertino, St., 229, **231**, 232
Journal of Psychological Type, The, 188–189
Judas Iscariot, 224, **225**
judicial astrology, 163
Julia sets, 102–103, 110–111
Jung, Carl, 80, 186–187, **187**, 196
Jupiter, 201

K

Kármán line, 3, 4
Kavanaugh, Brett, 210
Kaysing, Bill, 246–248, 252
Kean, Leslie, 87
Kelly, Ivan, 169–170
Kelly, Scott, 85
Kennedy, John F., 83, 245, **245**
kephalonomancy, 51
Kessler, Ronald, **216**
Key to World Events, The, 46
K'iche' Maya, 140
Kindred Spirits, 118
King, Larry, 152
King Kong, 135
Kirby, William, 131
knocking on wood, 220–221
Koecher, Hana, **216**

Koecher, Karl, 216, **216**
Kogan, Aleksandr, 195–196
Kosinski, Michal, 195–196
kosmotechnische Weltanschauung (astronomy of the invisible), 41
Kraft, Alexander, 27
kraken, 133
KrioRus, **158**
Kubrick, Christiane, 243
Kubrick, Stanley, 241–244, **243**, 247
Kusche, Larry, 100

L

Lair, Pierre-Aimé, 18
Lakshmi (goddess), 59
Larson, John, 212
Last Supper, 224, **225**
Lavery, Henry C., 73
Lawson, Alfred William, 265–274, **266**, **273**
Lawson Aircraft Corporation, 268
Lawson Money System, 268
Lawsonomy, 265–274
laymen, definition of, 181
Leader, 14–15
Leary, Timothy, 152
Led Zeppelin, 108
lenticular clouds, 79
Leo (astrological sign), 165
Leonardo da Vinci, 37
lesther, 272
levitation, 228–240
Lewes, George, 14–15
Ley, Willy, 46, 49
Ley Hunter, The, 179, 181
Ley Hunter's Manual (Watkins), 175
ley lines, 171–182
ley men, 173–174, 181
Liber monstrorum, 130

Libra (astrological sign), 164, 165
lie detector tests, **ix**, 209–218
Lieder, Nancy, 143–144
Lilienfeld, Scott, viii
Limbo of the Lost (Spencer), 95–96
Lincoln, Abraham, 124
Lincoln, Mary Todd, 115, **115**, 124
Lincoln, Willie, 115
Lindemann, Michael, 111
Lindsay, James (Lord Lindsay), 235–236
liquid nitrogen, **155**
lituus (diviner's staff), 201–202, **201**
Loch Ness monster, 128, **128**, 134–135
Loeb, Avi, 86
Loki, 224–225
Long Man of Wilmington., 173, 174
Lost Art of Reading Nature's Signs, The (Gooley), 207
Lou Gehrig's disease, 61
Louis XIII, 284
Loveland Frogman, 128, **129**
Lovelock, James, 107
lucky horseshoes, 223
Lukens, Isaiah, 33, 34
lunar cycles, 163
lunar eclipses, 7–8
lung mei energy lines, 178
lung-gom-pa runners, 233–234

M

M (crop circle witness), 103, 110–111
Mace, Nancy, 87–88
Magic and Mystery in Tibet (David-Neel), 233–234
magnetic current, 178
magnetic declination, 99
magnetism, Lawson and, 272

Magnus, Albertus, 130
Makemson, Maud Worcester, 141
Mani the parakeet, 204
Manlife (Lawson), 267
Manual of Cheirosophy, A (Heron-Allen), 56
Marine Sulphur Queen, 94–95, 101
marshmallow experiment, 61
Marston, William Moulton, 210, 211–212, 218
Martin PBM Mariner flying boat, 91–92
Maya, The (Coe), 141
Maya apocalypse. *See* 2012 phenomenon
Maya Long Count calendar, 138–139, 140
Maya people, 147–148, **148**
Maya(n) apocalypse, 137, 138–139, 141–143
Mayan Factor, The (Argüelles), 142
Mayanism, 141–142, 147–148
McKenna, Terence, 141, 142, **142**
Meaden, Terence, 107
meditative breathing, 234
Meet the Parents, 210, **210**
Melampus, 56
melanoma, 54
"Menorgs," 272–273
mesmerism, 73
Mesoamerican Long Count calendar, 138–139
mesomorphs, 192
methane, 262, **262**, 263
methane hydrates, 98
Mexico Mystique (Waters), 142
Miami Herald, 95
Michell, John, 176–178, **178**, 180
Midnight Liner, 268

Milky Way, 41
Mitchell, David, 55
moleosophy, 56
moles/beauty marks, 56
Mongolian death worm, 128
Monroe, Marilyn, 83
Montes, Ana, 216
Moody, Blair, 273
moon
　landing on, 241–252
　lunar cycles, 163
　lunar eclipses, 7–8
　rocks from, 244, **244**
　World Ice Theory and, 41, 42–44, **43**
Moore, Patrick, 104
mota gas, 26–27
Mothman, 128, **129**, 130
Mu (alleged missing continent), 146
Mumler, William, 115–116, 124
Murray, T. Patrick, 242–243
Myers, Clarence "Chief," 186
Myers, Isabel Briggs, 185–187, **186**
Myers, Katharine, 187
Myers-Briggs Type Indicator (MBTI) test, 184–189, 194

N

NASA
　flat Earth and, 8, 9, 10
　magnetic pole reversal and, 145, 146–147
　moon landing and, 241–242, 246–247, 248, 251, 252
　UFOs/UFOlogy and, 77, 84, 89
　zodiac signs and, 167–168
natal charts, 165, 166, 167, 169
National Academy of Sciences (NAS), 218
National Black Cat Day, 222
National Geospatial Intelligence Agency, 87
National Institute of Anthropological History (Mexico), 147
National Oceanic and Atmospheric Administration (NOAA), 254
National Reconnaissance Office, 87
National Security Agency, 214
National Socialists, 48–49
National Trail (UK), 174
nationalistic socialism, rise of, 46
natural astrology, 163–164
Nature, 169
navigation, birds and, 207
"Navigation Problem No. 1," 90
Nazca Lines, 176–177, **177**
neurofibromatosis, 54
neuropreservation, 153, 158
Neuroterus valhalla, 133–134
New Age movement, 179–180
New Scientist, 104
New York Times, viii, 21, 83–84, 87, 137
News Nation, 88
Newton's laws, 266–267
Nibiru (alleged planet), 143, 145
Nietzsche, Friedrich, 136
"Night Hag, The," 121
Nimitz, USS, 78
Nimoy, Leonard, 127
Ningen, 128, **129**
Nix, Alexander, 196
Nixon administration, 247
Norgay, Tenzing, 135
north, true versus magnetic, 98–99
Nosek, Luke, 152, **158**
NTSC format, 251
nuclear-powered cars, 29–30, **29**
numbers, unlucky, 223–225
numerology, 224–225

O

Observer, 282
Occult Science in India and Among the Ancients (Jacolliot), 233
OCEAN acronym, 193–195, **193**
ocean currents, energy from, 35
ocean floor topography, 99
O'Connor, Sinéad, 89
octopus, giant, **133**
Odyssey, The (Homer), 200
Office of the Director of National Intelligence, 84, 85
Ohtsuki, Yoshi-Hiko, 107
Oklahoma Octopus, 128
Old Bedford River, 6, 7
Old Straight Track, The (Watkins), 172, 176
Olmecs, 139
On the Constitution of the Universe and of Man (Ideler), 191
On the Track of Unknown Animals (Heuvelmans), 131
onychomancy, 56
Operation Independence, 282
Ophiuchus, 167–168, **168**
Ordnance Survey maps, 171
ornithomancy, 199, 200–202. *See also* auguries
osmotic shock, 156
Otis Elevator, 225
Ouija boards, 283
Ovilus, 118–119
oxygen, 17, **17**

P

palmistry/palm reading, 57–62, **57**
Parade Magazine, 161
parakeets, 204
Parallax (Samuel Rowbotham), 6
paranormal investigators, 117–118. *See also* ghosts/ghost hunting
Paranormal State, 118
paraskevidekatriaphobia, 225
pareidolia, 120
parhelia, 77, **82**
parrot fortune tellers, 204
Patterson, Roger, 126–127
Paul the Octopus, 204
Peczely, Ignaz von, 53–54
Pedley, George, 105
Pennsylvania State League (baseball), 267
Pensacola News Journal, **77**
perpetual motion machines, 31–39, **32**, **35**, **38**, **39**
personality psychology, 183–196
Pew Research surveys, 255
Philadelphia Gazette, 32
Philip Morris, 192
Philosophical Transactions of the Royal Society of London, 13
Phrenological Museum, 67
phrenology, 51, 64–74, **68**, **72**
phrenomagnetism, 73
physiognomy, **69**, 70
pillimancy, 51
Pinkerton detective agency, 24–25
piracy, 99
Pisces (astrological sign), 165, 169
pixie rings, 104, **104**
Pizzagate, 10
Planet Nibiru collision course theory, 143–144
planets, astrology and, 165–166, 168
plasma vortices, 107
platypuses, 132
Pliny the Elder, 133
Plot, Robert, 104
podomancy, 57
polar bears, 255, **255**
pole reversals, 143, 145, 146
Poliakoff, Martyn, **39**
polygraph, 209–218
Popol Vuh, 140
Popular Science, **35**, 38
Preacher Bat Boy, 128
Priam, 200
Priestley, Joseph, **17**
Princeton, USS, 78
procession, 169
Project Blue Book, 80
Prometheus, 199, **199**
pseudoscience, definition of, vii
Psychologische Typen (Jung), 186
psychometric data, 195
Psycograph, 73–74, **74**
Ptolemy, 40, 162
Publius Claudius, 203, 205
pullarii, 202–203
Punic Wars, 203, 205
"pyrotron," 20

Q

Quasar, Gian J., 92
Quigley, Joan, 161–162, **162**, 170
quizzes, personality, 183–184, **184**

R

R. J. Reynolds, 192
race and racism, 70, 215
Radford, Benjamin, 103
Ragnarok, 47
Ragnarok: The Age of Fire and Gravel (Donnelly), 47
Rahu, **166**
random points theory, 182
Raymond, R. W., 282
Reagan, Nancy, 161–162, 170
Reagan, Ronald, 160–162, **161**, 170
Redheffer, Charles, 31–34, **32**
Reeser, Mary, 20
refraction, 9
Reid, Harry, 84
Reid, John, 212
relativity, 49
Remus, 197–198
Renaissance, vii
Revelation, Book of, 224
Richelieu, Cardinal, 284
Ridgeway, The (trackway), 174–175
Ridgway, Gary, 216, **216**
Rocketdyne, 246
Rocketman: Mad Mike's Mission to Prove the Flat Earth, 2–3, 11
rockets, 2–4
"rockoon," 4, 11
Rockwell, Norman, **35**, 38
Rogan, Joe, 88, 243
Rolli, Paul, 13
Romani culture, 57
Romulus, 197–198
roosters, using for prediction, 51
Rosalie, 93, 101
Roswell, New Mexico, 79, 80, **81**
Rowbotham, Samuel (Parallax), 5–6
rumpology, 51–53
"Russian hail," 79

S

Sacalxot, Martin, 147–148
Sagan, Carl, vi, ix, **x**, 81, 83

Sagittarius (astrological sign), **164**, 165, 168
Saint Teresa of Spain (Colvill), 277
Samudrika, 59
Samuel, 112–113
Sanderson, Ivan T., 96–97, 131
Sandy Hook school shooting, 10
Sasquatch (Bigfoot), 125–128, **127**, 136
Saturn V rockets, 246
scatomancy, 51
Scientific American, 36
Scientific Opinion, 6
scientific process/method, vi–vii, 123
Scorpio (astrological sign), 165, 168, **168**
Scorpion (submarine), 96
Scot, Michael, 58–59
séances, 115, 124, 235–238
seizures, 232
selective attention, 121
Sellers, Coleman, 33
Sellers, Nathan, 33
Senate Intelligence Committee, 84
Senate Judiciary Committee, **195**
Senate's Small Business Committee, 273
SETI (Search for Extraterrestrial Intelligence) committee, 82, 86
sexegenary calendar (ganzhi), 167
Shani (deity), 59
Shaw, George, 132
Sheepsquatch, 128
Sheldon, William H., 192
Shepard, Alan, **251**
Shooting Stanley Kubrick, 242
Sibrel, Bart, 247–248, 252

Signs, 103
Sitchin, Zecharia, 143
Skeptical Inquirer, 103
skin, using for prediction, 55–56, **55**
skin galvanic resistance, 212
Slaight, Jim, 78
sleep paralysis, 121
slit lamps, 53
Slocum, Joshua, 93–94, 101
"Sludge the Medium" (Browning), 237
Snarly Yow, 128
Snowden, Edward, 258
social media, 195–196
Société d'Autopsie Mutuelle (Society of Mutual Autopsy), 66
solar irradiance, 261
solar maximums, 145–147
space race, 245–246. *See also* moon landings, fake
Spalding, Tim, 56
Spencer, John Wallace, 95
spies, 216, **216**
spirit photography, 115–116, 124
spirit telephones, 115
Spiritualism, 114, 234–235
Spitzka, Edward, 67
Splash, 135
spontaneous human combustion, 12–20
Spray, 94, 101
Spurling, Chris, **128**
Spurzheim, J. G., 68
squids, giant, 132–133, 135
SSTV (slow-scan television), 251
St. Elmo's fire, 79
Stakes, Waldo, 3, 11
Stallone, Jackie, 51, **51**, 52
Steiner, Don. R., 282
Stone, Sharon, 210, **214**
Stonehenge, 102–103, 107
Straight Track Club, 175

Strategic Communication Laboratories (SCL), 195–196
strength-to-size ratios, 240
suction and pressure, theory of, 265–266, 271–273. *See also* Lawsonomy
"sun dogs," 77, 79, **82**
sun signs, 165, 166
superstitions, 219–226
swamp gas, 82
sweating, lie detection and, 212, **217**
swimming a witch, 211

T

Taurus (astrological sign), 165
Taylor, Charles C., 90–91, 100
Technisches Museum, 39
Teissier, Elizabeth, 168
temperance movement, 18–19
temperature drops, 118, 119, 120
temporal-lobe seizures, 232
Teresa, St., 277
Teresa of Ávila, St., 229–230, **230**, 232
Tesla, Nikola, 38
thermodynamics, laws of, 35–37, 38, 122
Thiel, Peter, 152, **158**
Thomas, Henry, 20
Tice, Russell, 214
"Tiggy, Tiggy Touchwood," 220
time crystals, 36
time warps, 95
Times of London, 93, 101, 237
tobacco industry, 192–193, **256**, 257, 259
"tocca ferro," 221
Tonight Show with Jay Leno, The, 28
traits versus types, 189
transcendental consciousness, 180

• INDEX | **307**

transhumanists, 157
trials by ordeal, 211, **211**
triskaidekaphobia, 225
Truman, Harry, 83
Trump, Donald, 196, 252
tsunamis, 98
Tumulty, Karen, 161
Twain, Mark, 60
2001: A Space Odyssey, 241, 242, 243, 247
2012, 143
2012 phenomenon, 137–148
type A personality, 191–193
Tyson, Neil deGrasse, 81
Tzolk'in calendar, 139–140

U

UAP Independent Study Team, 89
UAP Task Force, 87
UAPs (unidentified anomalous [or aerial] phenomena), 79, **80**, 84–85, 87–89
UFOs/UFOlogy, 76–89, 95–96, 105–106, **105**, 107–108, 176–177. *See also* aliens/extraterrestrial life
United Nations Conference of the Parties (COP) climate summit, 263
United Nations logo, 10, **10**
United States v. Scheffer, 216
Universal Zetetic Society, 6–7
University of Des Moines, 270
University of Lawsonomy, 270–271, **271**, 273–274
unlucky numbers, 223–225
Urban VIII, Pope, 232
US Coast Guard, 100
US Patent Office, 38

V

Valhalla, 224
"valueless money," 268
Van Allen radiation belt, 247–248
Vedic texts, 57
Vegetable Man, 128
Velásquez García, Erik, 147
Viet Cong tunnels, 282
Vietnam War, 247, 282
View from Atlantis, The (Michell), 178
Virginian Bunnyman, 128
Virgo (astrological sign), 165, 168
viruses, 134
vitrification, 156
von Butlerov Alexander, 236
von Däniken, Erich, 176–177

W

Wagner, Rudolph, 66
Wallace, Alfred Russel, 6, 236
Wallace, Wilbur "Shorty," 125–126
Walters, Ed, 76–77, **77**
Warner, William John (Cheiro), 59–60, **60**
water, cryonics and, 155–156
Waters, Frank, 141–142, 143
Watkins, Alfred, 171–175, **172**, 176, 178, 180, 182
Watkins, Allen, 172
We Never Went to the Moon (Kaysing), 246–248
weather/weather patterns, 99, 107, 206–208, 222, 258–260, **263**. *See also* climate change denial
WEIRD acronym, 194–195
Weiss, Edmund, 48
Welles, Horatio, 67
Wells, Samuel, 71
Welteislehre. *See* World Ice Theory
Wessely, Christina, 45, 46
What Does Joan Say? (Quigley), 161
White, Gilbert, 131

Whitman, Walt, 66–67, **67**
wick effect, 19–20
Wilde, Oscar, 60
Wilkens, Henry, **213**
Willard, Clarence E., 238
Williams, Ted, 150–151, **151**, 157
Wilson, Woodrow, 24, 94
Wilson's disease, 54
Wired, 246
wishbones, 222
Wistar Institute of Anatomy and Biology, 67
witches/witchcraft, 211, **211**, 221, 225, 284
women, phrenology and, 70, 71
women's equality, 70
Wonder Woman, 209–210
Woolhope Naturalist and Field Club, 172
World Ice Theory, 40–49
World Wildlife Fund, 92
World's Fair, 74
Wylie, Christopher, **195**
Wyndham-Quin, Windham (Lord Adare), 235–236
Wynn, Charles, 236

Y

Yeti, 128, 135
Yoakum, Benjamin, 24–25, **24**
YouGov polls, 166

Z

Zetans, 143–144, **144**
ZetaTalk, 143
Zetetic Astronomy (Rowbotham), 5–6
Zeus, 199, 200
Zhongyuan Ghost Festival, **113**
zodiac signs, 164–165, **164**, 167–169. *See also* astrology

Photo Credits

Alamy: NASA Photo pp. x, 8; SPCOLLECTION pp. 1 (bottom), 65; Heritage Image Partnership Ltd. p. 5; John Worrall p. 6; Charles Walker Collection pp. 12, 60, 168, 235; Chronicle p. 15, 97, 230, 276, 278, 280; Edwards Online Media p. 16; Mathilde Receveur p. 25 (left); Maurice Savage p. 25 (right); Allan Cash Picture Library p. 27; Rachel Royse p. 29; Timewatch Images p. 31; Science Photo Library p. 37; Science History Images, pp. 50, 55, 190; Lebrecht Music & Arts p. 51; Gameover p. 66; Michelle Bridges p. 69; VintageMedStock p. 72; MasPix p. 81; Lars Johansson p. 82; Stephen Saks Photography p. 83; Arthur Greenberg p. 86; CBW p. 91; Peter Hermes Furian p. 92; Penta Springs Limited p. 93; US Department of Defense Archive p. 94; Frank Hecker p. 104; James Thew p. 105; Robertharding p. 106; Christopher Jones p. 108; David Wootton p. 110; Imaginechina Limited p. 113; History and Art Collection p. 115 (top); Allstar Picture Library Limited p. 117; Alistair Heap p. 120; Michael Ventura p. 123; Zip Lexing p. 130; The History Collection p. 134; YAY Media AS p. 139; Imago p. 148; Victor Habbick Visions/Science Photo Library p. 152 (top); CPA Media Pte Ltd. p. 166; Tony Peacock p. 171; Album pp. 172, 210, 243; Album/Giambettino Cignaroli pp. 227 (top), 231; Alto Vintage Images p. 173; SJBooks p. 175; DPA Picture Alliance p. 177; Emma Stoner p. 178; World History Archive p. 184; INTERFOTO pp. 187, 205; Kristoffer Tripplaar p. 195; The Picture Art Collection pp. 199 (top), 234; Pictures Now p. 199 (bottom); Tim Gainey p. 204; ZUMA Press, Inc. pp. 216 (top), 263; Life On White p. 219; Cagan Niron p. 221 (top); Pegaz p. 221 (bottom); Marina Pissarova p. 228; Mira p. 233; Bill Crump p. 244; Felix Mizioznikov pp. 227 (bottom), 253; Tom Walker p. 255; Shawshots p. 256; Geoff Smith p. 258; fStop Images GmbH p. 262; Classic Image p. 279.

Adobe Stock: YouraPechkin (Generated with AI) p. 38; MiaStendal (Generated with AI) p. 47; i-element (Generated with AI) pp. 75 (top), 96; Michael (Generated with AI) p. 125; FIFTY8 Design Studio (Generated with AI) p. 197.

AP Images: Eric Risberg p. 162; *Houston Chronicle* p. 245.

Everett Collection: © TriStar p. 214.

Freepik: Mixzer (Generated with AI) p. v

Getty Images: Pictorial Parade/Archive Photos p. ix; Hulton Archive/Stringer p. 17; Bettmann pp. 19, 127, 213, 249, 275; Kirn Vintage Stock/Corbis Historical p. 21; Werner Cohnitz/ullstein bild pp. 41, 48; Cokada/E+ p. 43; Hulton Deutsch/Corbis Historical p. 57; Ed Bock/The Image Bank p. 61; Universal History Archive/Universal Images Group Editorial p. 64; National Archives/Stocktrek Images p. 67; Photo 12/Universal Images Group pp. 68, 283; Ron Miller/Stocktrek Images pp. 75 (bottom), 141; ktsimage/iStock p. 76; Shawn van Eeden/500Px Plus p. 80; Tom Williams/CQ-Roll Call, Inc. p. 88; Victor Habbick Visions/Science Photo Library p. 90; Heritage Images/Hulton Archive p. 102; London Stereoscopic Company/Stringer/Hulton Archive p. 116; Scotspencer/E+ p. 119; Keystone/Stringer/Hulton Archive p. 128; AFP/Stringer pp. 137, 251; DEA/ICAS94/De Agostini pp. 149 (top), 197; Malte Mueller/fStop pp. 149 (bottom), 164; General Photographic Agency/Stringer/Hulton Archive p. 152 (bottom); Choja/E+ p. 155; Larry Armstrong *Los Angeles Times* p. 157; Bloomberg p. 158 (left); Kumarworks/DigitalVision Vectors p. 160; Dirck

Halstead/Hulton Archive p. 161; Nastasic/DigitalVision Vectors p. 201; CBS Photo Archive p. 209; Fototeca Storica Nazionale/Hulton Archive p. 211; Peter Dazeley/The Image Bank p. 217; CSA-Printstock/DigitalVision Vectors p. 222; Malerapaso/iStock p. 223; Traveler1116/DigitalVision Vectors p. 225; Oonal/iStock p. 226; Visual Studies Workshop/Archive Photos p. 240; Corbis Historical p. 248; HUM Images/Universal Images Group Editorial p. 250 (top); George Rinhart/Corbis Historical p. 266.

Lincoln Financial Foundation Collection: William Mumler p. 115 (bottom).

NASA: JPL/USGS p. 1 (top); Michael Collins p. 241; JSC p. 250 (bottom).

Shutterstock: Marti Bug Catcher p. viii; Elena Schweitzer p. 2; Paul Buck/EPA-EFE p. 3; Mr_seng p. 40; Magic Panda p. 52; VectorMine p. 54; LMPark Photos p. 62; Shutterstock p. 109; Alba Foto p. 112; JM-MEDIA p. 129 (bottom); LianeM p. 138; Limbitech p. 144; Citrina p. 150; Asmus Koefoed p. 154; Al.geba p. 163; Viktoriia_Patapova p. 183; Badproject p. 185; Julee Ashmead p. 193; Freer p. 203.

Science Photo Library: GUSTOIMAGES p. 180

The Royal Society: David Edward Hugh Jones p. 39

USA Today: © *Pensacola News Journal* p. 77

U.S. Navy: p. 78

Public Domain: The American Museum Journal, americanmuseumjo15amer p. 23; Charles Redheffer, 1812 {{PD-US}} p. 32; Library of Congress, LC-USZ62-20997 p. 34; Norman Rockwell; *Popular Science*, Vol. 97, No. 4 (October 1920), Bonnier corp. New York, ISSN 0161-7370 p. 35; Library of Congress, LC-H2- B-4880 p. 74; Dawn Keetley, Lehigh University p. 103; Gibson, J. (1887), *Monsters of the Sea: Legendary and Authentic*, Thomas Nelson and Sons, London {{PD-US}} p. 133; Apex Photo Company, 1939 p. 151; Courtesy of Katharine Myers p. 186 (left); Herbert Rose Barraud (1845–ca. 1896) p. 237 (left); Library of Congress, LC-USZ62-94791 p. 237 (right); A.F.B. Geille after J. Boilly. Wellcome Collection. p. 254; C-SPAN p. 261; Library of Congress, LC-DIG-hec-19510 p. 265; Wisconsin Historical Society p. 269; Alfred Lawson/Cosmo Power Company Publishers p. 273.

Wikimedia

The following images are used under a Creative Commons Attribution CC BY-SA 4.0 License (creativecommons.org/licenses/by-sa/4.0/deed.en) and belong to the following Wikimedia Common users: United Nations p. 10; p. 126; Tim Bertelink p. 129 (top right); Ronald Kessler p. 216 (bottom); *Manlife* documentary/dir. by Ryan Sarnowski and Susan Kerns p. 271.

The following images/ are used under a Creative Commons Attribution CC BY-SA 3.0 License (creativecommons.org/licenses/by-sa/3.0/deed.en) and belong to the following Wikimedia Common users: Jon Hanna p. 142.

The following images are used under a Creative Commons Attribution CC BY-SA 2.0 License (creativecommons.org/licenses/by/2.0/deed.en) and belong to the following Wikimedia Common users: Heisenberg Media/Image by Dan Taylor p. 158 (right).

The following images are used under a Creative Commons Attribution CC BY-SA 1.0 License (creativecommons.org/licenses/by/1.0/deed.en) and belong to the following Wikimedia Common users: Pieter0024 p. 129 (top left).

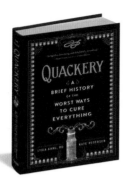 

Lydia Kang, MD, and Nate Pedersen are also the authors of *Quackery: A Brief History of the Worst Ways to Cure Everything* and *Patient Zero: A Curious History of the World's Worst Diseases*.

ABOUT THE AUTHORS

PHOTO COURTESY OF THE AUTHOR

LYDIA KANG, MD, is an associate professor of internal medicine and author of nonfiction books, essays, and adult and young adult fiction. Her novels include *Opium and Absinthe, A Beautiful Poison, The Half-Life of Ruby Fielding,* and *The November Girl.* Her essays have appeared in *Flatwater Free Press* and *JAMA.* She is also the author of the best-selling Star Wars novel, *Cataclysm.*

WILL FOERSTER

NATE PEDERSEN is a librarian, historian, and freelance writer in Savannah, Georgia. His nonfiction has appeared in many national outlets and he is a regular contributor to *Fine Books & Collections* magazine. He edited the Lovecraftian anthologies *The Starry Wisdom Library* and *The Dagon Collection.* His website is http://natepedersen.com.